ELECTRONICS COMPUTER AIDED DESIGN

Edited by P. L. Jones *and* Anne Buckley

D1739407

Manchester University Press

Manchester and New York

Distributed exclusively in the USA and Canada by *St. Martin's Press*

Published by Manchester University Press
Oxford Road, Manchester M13 9PL, UK
and Room 400, 175 Fifth Avenue, New York, NY 10010, USA

Distributed exclusively in the USA and Canada
by St. Martin's Press, Inc.,
175 Fifth Avenue, New York, NY 10010, USA

British Library cataloguing in publication data applied for
Electronics computer aided design
  1. Electronic equipment. Design. Application of computer
systems
I. Jones, P. L. II. International journal of electrical
engineering education
621.381'042'0285
ISBN 0-7190-3079-X

Library of Congress cataloging in publication data applied for

ISBN 0-7190-3079-X

This is a special edition in book form of
The International Journal of Electrical Engineering Education, Vol. 26 nos 1 & 2

Printed in Great Britain
by H Charlesworth & Co Ltd, Huddersfield

CONTENTS

FOREWORD

Computing is now an integral element of engineering pursuit and nowhere more so than in the design and testing of integrated circuits. Software to produce complex digital and analogue integrated circuits is itself very complex and multi-faceted. Designers need to capture abstract ideas to model their system, prepare schematics, simulate at many levels of abstraction, create test data, layout the chip, generate manufacturing data for fabrication and finally thoroughly test the chip. This activity is now so germane to an electronics education that all institutions responsible for producing electronic engineers need IC design as part of the engineering appreciation process.

The software complexity and specialism makes new demands on both the support mechanisms and the educational process. Much care must be exercised in producing the right material for students of widely differing educational standards. Support in the laboratory and of the software tools has to be of the highest calibre. Software designed to the highest professional standards is an essential starting point and students must be exposed to commercially relevant and timely products.

The UK Higher Education Electronics Computer Aided Design (ECAD) Initiative has attempted to meet some of these exacting requirements. It has brought unparalleled cooperation at many levels of involvement. The establishment of the scheme involved the University Grants Committee, the Science and Engineering Research Council, the National Advisory Board for Polytechnics and Colleges, the Department of Trade and Industry and the Computer Board. Polytechnics and colleges of higher education link with UK universities. Suppliers and academics interact in harmony (mostly). A large community of people is now working together towards better education in this important area; a newsletter is published, a national user group has formed and many supplier-oriented user groups meet regularly.

In producing this book, academics in the UK have presented their experiences and activities in teaching IC design for the first time in a single volume. Papers address topics across the spectrum of technology options. Twenty-six papers are published from twenty or so institutions. It is an unrivalled opportunity for educators in the UK and world-wide to share the excitement and traumas of a new teaching process.

I cannot let this chance pass without expressing my thanks to the publishers and the editors for putting together this important contribution to the literature. Furthermore, we must not forget the frequently unsung efforts of all the staff who, behind the scenes, make everything work smoothly: the management committee members of the ECAD Initiative, lead site staff, all the support staff of the academic institutions and the suppliers and everyone involved in the user group and newsletter activities. Finally I must express my thanks, on behalf of all the ECAD community, to Peter Jones of Manchester University, the executive officer of the Initiative, and David Boyd, the secretary of the committee, who between them undertake an incredible amount of work to make the Initiative such a success.

Professor ERIK DAGLESS
University of Bristol, Chairman of the ECAD Management Committee

THE UK HIGHER EDUCATION ELECTRONICS COMPUTER AIDED DESIGN INITIATIVE

P. L. JONES
Department of Electrical Engineering, University of Manchester, England

1 INTRODUCTION

In 1986, funds totalling £8M were allocated by the University Grants Committee (UGC), the Department of Trade & Industry (DTI), and the Science and Engineering Research Council (SERC), to provide industrial standards in software and hardware for teaching electronics computer aided design (ECAD) in degree courses throughout the United Kingdom. The origins of this initiative go back to 1984 with the establishment of working parties representing the interests of universities and polytechnics to explore means for introducing integrated circuit design into the curriculum of the majority of undergraduate courses in electronics.

Thus, following 15 months of negotiation and debate, the first positive steps were taken in late 1985 when the UGC and SERC agreed jointly that funds would be made available for purchasing CAD packages for microelectronics teaching and research in the university sector. A combined SERC–University panel was therefore asked to make the technical assessment both of the needs of institutions and of the availability of appropriate software packages.

About 20 companies were invited to tender. It was clear from the outset that no single CAD supplier could hope to satisfy the full spectrum of requirements and the panel was therefore faced with the difficult task of obtaining a best fit to the specification, while remaining within the constraints of a limited budget. A survey of university hardware needs had shown that at least 300 seats were required for ECAD. Since CAD packages were traditionally sold on a per seat basis, the cost of only modest capability appeared to be in the region of £30M. The panel's policy therefore was to negotiate a single licence for educational use throughout the U.K. at a cost per package proportional to the importance of the purchase in meeting the universities' CAD requirements and to the expected level of take-up, rather than to negotiate a cost per seat or per site.

It proved relatively easy to convince potential suppliers that it made commercial sense to come in with a low bid to capture the educational market. They were also persuaded that annual support charges could be low, provided a system of 'lead sites' was operated. The suppliers thus agreed to provide full industrial support for a small number of lead sites who in turn were supporting many user sites, none of whom have direct contact with suppliers. It was also agreed that suppliers should include in their bids a charge to extend the deal to include polytechnics and other higher education institutions in the public sector administered by the National Advisory Board (NAB).

Following acceptance of the SERC–University panel recommendations, university establishments were informed of the developments and were invited to join a scheme designed to provide technical support and maintenance for the software purchased. Shortly afterwards, the DTI provided the funds necessary for the public sector institutions to be included and also for the purchase of additional software. Thus, the aim to achieve blanket licences for the educational use of professional software for ECAD in teaching and research was attained and work could begin on the establishment of a management and support infrastructure.

The terms of reference and constitution of an ECAD Management Committee were agreed by the co-sponsoring organisations UGC, DTI, NAB and SERC. The committee met for the first time in December 1986, taking full responsibility for monitoring and progression of the ECAD Initiative. Lead sites established at the Universities of Manchester, Newcastle and Surrey and at the Polytechnics of Huddersfield and Middlesex are now entering their third year of operation, providing, with two industrially-based lead sites, support for the broad range of packages described in Appendix A to this paper. The administrative arrangements described in Appendix B are entirely self-financing on a recurrent basis through income from membership subscriptions from the 90 educational establishments currently participating in the scheme. Additional capital for purchase of new software is provided by the co-sponsoring organisations.

2 DISCUSSION

Although the ECAD Initiative currently has a sound financial base and management structure, its future stability and growth depends strongly on the continued support from members. The Management Committee's remit must therefore be to provide clear value for money for the member institutions, bearing in mind that for many departments, the ECAD membership fee represents a significant fraction of their recurrent budgets. Thus, raising the standard of microelectronics design capability through the provision of state-of-the-art ECAD tools is only one aspect to be considered. The cost of maintaining each of the tools needs to be assessed continually in educational terms and, where necessary, the current software portfolio must be enhanced or pruned to suit changing needs. There is also a duty to promote awareness of the importance of ECAD in postgraduate and undergraduate training.

Over the next eighteen months, a critical assessment will be made from data derived from questionnaires and from user groups to determine which of the current elements in the software portfolio be retained and what new elements need to be incorporated. Although the level of uptake of a package is one factor to be considered in assessing its educational value, it will not be the sole criterion. Equally important is the quality of work undertaken with a package and the relevance of that work to the training of an electronic engineering graduate. Research needs must also be taken into account in any assessment. Progress monitoring has already become an integral part of the ECAD Initiative management task. Within their first year of operation the lead sites underwent a major review, the outcome of which should lead to more effective user support in future. In addition, since in the initial software procurement special emphasis was placed on the provision of tools leading to silicon fabrication, a scheme has been introduced to subsidise to a level of 50% the cost of chips made for educational purposes. For the near future, a proposal to subsidise the purchase of IC test equipment, linked to existing CAD tools, is also under discussion. Clearly, every effort is being made to enhance the support of many different aspects of microelectronics design education.

3 CONCLUSIONS

All higher education institutions which offer degree courses with an electronic engineering content now have, through membership of the ECAD Initiative, a free licence to use a broad spectrum of modern industry-standard tools for the computer aided design of electronic systems. The breadth of the ECAD Initiative software portfolio is aimed at encouraging equally all aspects of educational needs, from high-level system behavioural specification to low-level circuit synthesis and analysis. Thus, with the variety of specialist interests within departments in the 90 member institutions, the diverse skills needs in the graduate intake to the electronics industry should be well satisfied for the foreseeable future[1,2].

4 REFERENCES

[1]   Jones, P. L., 'The Status of ECAD in Higher Education', *Journal of Semicustom ICs*, **5**, No. 2, pp. 39–40 (1987)
[2]   Pritchard, T. I., 'The U.K. Electronics Computer-Aided Design Initiative Progress Report', *ibid.*, **5**, No. 4, pp. 43–44 (1988)

5 APPENDIX A The ECAD Software portfolio 1986–88

5.1 *SILVAR LISCO*

Design Entry
SDS, CASS, EARS, I-SPICE, I-HILO
Database (SDS) and schematic entry system (CASS) supporting hierarchical design.
Routines (EARS) to extract data from database for other software packages.
Specific purpose interface tools (I-SPICE, I-HILO).
Simulation
HELIX, BIMOS, ANDI, SWAP, LOGAN, PPRG
HELIX provides a hierarchical hardware description language to describe behaviour with mixed gate level and system simulation capability. BIMOS is a logic simulator. ANDI is a mixed analogue–digital simulator. SWAP a switched capacitor simulator. LOGAN provides graphical analysis of simulation results from HELIX. PPRG performs a similar task to display output from BIMOS, ANDI and SWAP.

Layout
GARDS, CALMP, PRINCESS
General purpose gate array layout tools, (GARDS). A general purpose standard-cell layout tool (CALMP). A full custom VLSI layout editor (PRINCESS).
Libraries
HELIX libraries for TTL, ECL 10k, CMOS4000.
HELIX libraries for Texas Instruments Gate Arrays.
GARDS libraries for Texas Instruments Gate Arrays.
GARDS libraries for MCE Gate Arrays.

5.2 *GENRAD*
Simulation
HILO-3
Functional and gate level simulation. Automatic test pattern generation and fault simulation.
Libraries
A full range of standard part libraries is provided by Genrad. Libraries for Texas Instruments gate-arrays, MCE gate-arrays and for Silicon Micro systems 3 micron standard cells are provided by SERC.

5.3 *PRAXIS*
Simulation
ELLA
Hierarchical hardware and behavioural description language with capability for mixed system level and gate level simulation. Links to silicon technology.

5.4 *SILICON MICROSYSTEMS (SMS)*
Cell layout library
Library of standard cells in 3 μm CMOS technology. Compatible with both Silvar Lisco CALMP, MCE, BXCELLS, and Racal Redac ISIS. Limited process independence is achieved through use of envelope design rules for a range of silicon suppliers.

5.5 *RACAL REDAC*
VLSI design system
ISIS
Includes schematic capture, hardware description language with mixed level simulation capability, artifact based layout isomorphic with HDL, floor planning tools, auto-routing tools, and on-line design rule checking. It supports the SMS standard cell library and has an ELLA interface.
Printed Circuit Board Layout
REDCAD
Includes schematic capture, parts libraries and autolayout tools for the design of printed circuit boards on a low-cost configuration of an IBM-PC. The output interfaces to standard photo-plotter and pen-plotter formats.

5.6 *MICRO CIRCUIT ENGINEERING (MCE)*
Gate array design suite
BXDESIGN
BXDESIGN is available with schematic entry on IBM PC-XT or AT, on Apollo and, with text entry on VAX. It provides design capture and simulation for a family of CMOS gate-arrays.
BXLAYOUT is available for Apollo and allows gate-array layout to be completed.
Cell based design suite
BXCELLS
BXCELLS runs on IBM PC XT/AT or on Apollo. It generates a netlist for 3 μm standard cells based on SMS library.

5.7 *QUDOS*
Gate Array Teaching System
MINICHIP
Minichip comprises a hardware description language, a simulator, manual layout, and connectivity check software for introductory design work on a Ferranti ULA using an Acorn BBC Microcomputer.
Gate array design suite
QUICKCHIP
Quickchip, a gate array design suite with text based hardware description, simulator and layout tools for Texas Instruments gate arrays using an Acorn Workstation.

5.8 *EUROPEAN SILICON STRUCTURES*
VSLI design system
ES2 SOLO 1000/1200
Solo 1000 is a fully integrated suite of IC design software using both Silicon compilation techniques and macro cells. Designs are taken from schematic or netlist entry, through simulation, to automatic place and route, post layout simulation, package choice and design validation. An extensive range of libraries allows for a flexible design approach incorporating both random logic compilation and macro cell routing, directed at providing a smooth route to full-custom fabrication. SOLO 1200 extends this capability to analogue cells.

6 APPENDIX B Administrative arrangements for the Higher Education ECAD Initiative

6.1 *Administration*
The scheme is administered by the University of Manchester Regional Computer Centre (UMRCC). The UMRCC ECAD Office is responsible for the distribution and collection of software agreements and for the distribution of documentation. The University of Manchester provides central services to assist with administration, issues invoices for membership fees and monitors software issues and licence obligations, reporting as required to the ECAD Management Committee.

The Management Committee comprises the principal parties involved in the programme, namely, UGC, NAB, DTI, and SERC. The committee advices the University of Manchester on the general running of the scheme, monitors the accounts and sets the annual membership fee.

6.2 *Software agreements*
The agreements with the suppliers provide for a national licence for the software for use within the Higher Education sector. The software may be used for educational purposes only, which covers undergraduate teaching, postgraduate research and in most cases post-experience short courses for industry. Any commercial use is expressly forbidden without the prior consent of the supplier. Each institution must complete a standard educational licence agreement with the supplier before any software will be released.

Software updates will be used normally at the beginning of the summer term ready for use in the following academic year. The issuing of updates at other times will only occur in special circumstances. The software is authorised via CPU identity numbers. Certain suppliers have agreed to accept computers into the scheme only at six monthly intervals. All users should be aware of this constraint when planning the purchase of new hardware for their ECAD systems.

6.3 *Software support*
The low-cost maintenance agreements negotiated with the software suppliers provide updates and support for a small number of lead sites serving the user community. The lead sites are responsible for distribution of software, collation of bug reports, assisting with installation and training (not

necessarily free of charge), distribution of updates, liaison with suppliers on behalf of users, etc. User support is thus achieved by allocating a proportion of the membership fee to lead sites. The balance of the fee covers payments to suppliers and the cost of administration by the Management Committee.

6.4 *Membership*
All member institutions must undertake to pay the University of Manchester an annual membership fee. Access to the software is permitted only on acceptance of the terms of the educational licence agreements, as administered by the University of Manchester, acting on behalf of the Management Committee.

The membership fee (currently £4381.50) provides access to the packages on any number or combination of computers and a set of documentation for each package. A single nominated contact person is required to interface with the administrative centre at UMRCC. User contacts should not exceed one per lead site. The ECAD Office at UMRCC maintains a mailing list of contacts, to whom the twice yearly ECAD Newsletter is circulated to promote the exchange of views and information between the Management Committee, the lead sites, the software suppliers and the membership.

ABSTRACTS–ENGLISH, FRENCH, GERMAN, SPANISH

The UK Higher Education Electronics Computer Aided Design Initiative
This paper describes the setting up and operation of a scheme to provide industrial standards in software and hardware for the teaching of electronics computer aided design (ECAD) in degree courses throughout the United Kingdom.

L'initiative ECAD (conception assistée par ordinateur en micro-électronique) dans l'enseignement supérieur en Grande-Bretagne
Cet article décrit la mise en place et le fonctionnement d'une initiative ayant pour but de fournir des standards industriels en logiciels et matériels à l'enseignement de la conception assistée par ordinateurs de l'électronique (ECAD) dans tous les cours supérieurs en Grande-Bretagne.

Die Hochschulbildungsinitiative für computergestützten Elektronikentwurf (ECAD) im Vereinigten Königreich von Grossbritannien und Nordirland
Die Arbeit beschreibt die Aufstellung und Handhabung eines Projektes zur Erstellung industrieller Software- und Hardwarerichtlinien für den Unterricht des computergestützten Elektronikentwurfs (ECAD) in akademischen Kursen im ganzen Vereinigten Königreich.

La iniciativa sobre la educación superior del ECAD en el Reino Unido
En este articulo se describe la puesta en marcha y el funcionamiento de un esquema de enseñanza, que proporcione conocimientos a nivel industrial, tanto en el software como en el hardware, en el área del diseño electrónico asistido por computador (ECAD), en los cursos de grado en todo el Reino Unido.

THE DEVELOPMENT OF AN UNDERGRADUATE CURRICULUM FOR VLSI DESIGN

R. E. MASSARA, A. D. P. GREEN, R. J. MACK and P. D. NOAKES
Department of Electronic Systems Engineering, University of Essex, England

1 INTRODUCTION

The Department of Electronic Systems Engineering at the University of Essex runs Bachelor of Engineering (B.Eng.) degree schemes in electronic engineering which place an emphasis, that we regard as distinctive, on the design and implementation of electronic systems. This emphasis corresponds closely to the theme of most of the Department's research activity. The emergence of CAD-based design methodologies, and of custom device styles, which make VLSI design accessible to the non-specialist system designer, very clearly creates new demands for the provision of appropriate programmes of theoretical and practical training for the electronic systems engineer. This was an area which the Department decided to include in its undergraduate programme some three years ago. The University of Essex was not, however, equipped with fabrication resources to support chip design activities in common with most institutions of higher education in the U.K., and, whilst well served in respect of general computational resources, possessed only research-scale CAD facilities to support integrated circuit design.

The emergence of the Electronics CAD (ECAD) initiative was, in this context, of very great interest to us at Essex. The aims of the initiative, generally, were to provide hardware and software support for engineering CAD. In the case of electronics, the areas addressed were integrated circuit (IC) and printed circuit board (PCB) design, and tools for gate- and higher-level circuit stimulation. A key objective of the initiative has been the provision of commercial IC design CAD tools which provide a cost-effective route to silicon, allowing UK undergraduate students to gain real experience in IC design and implementation.

The extension of our teaching programme in CAD to include VLSI design concepts coincided with a time in which the Department was also enhancing its established B.Sc. programmes to meet B.Eng. accreditation requirements. Since CAD experience has been identified in the Engineering Council's SARTOR document[1] as one of the important facets of an engineer's exposure to current-generation tools, we took the opportunity to review our activities in the CAD field completely. The resulting programme, which makes a significant contribution to our Engineering Applications (EA)* activities, depends markedly on hardware and software resources that were very largely obtained via the ECAD initiative.

2   FIRST AND SECOND YEAR ACTIVITIES

In the first and second years of our three-year B.Eng. scheme in Electronic Systems Engineering, all students take a common programme of courses irrespective of final-year optional specializations. During these two years, students take courses which, through two major course themes, establish a foundation on which final-year work in VLSI Design is based.

The first of these themes is that of Semiconductor Device Engineering, which is developed through the following courses:

1st year	Introduction to Electronic Circuit Design	(25 hours)
	Electrical Materials and Components	(20 hours)
2nd year	Semiconductor Devices	(15 hours)

Through these courses, students receive a grounding in:

● the physics of semiconductor materials and devices
● integrated circuit fabrication techniques (the PLANAR and related processes)
● characteristics of integrated circuit devices: passive components, BJT, FET and MOST
● device models for use in simulation program applications.

The second pertinent theme is that of Digital Systems Engineering, which is developed through the following courses:

1st year	Digital Systems Engineering	(25 hours)
2nd year	Digital System Design	(30 hours)

Aspects of these courses provide further support for the later work in VLSI, including:

● digital systems design and architecture
● alternative styles of implementation, including PLDs and Gate Arrays
● digital systems simulation at gate and function level
● testability in digital systems design.

In addition to these lecture courses, students undertake a number of supporting laboratory activities which make a significant contribution to the Department's programme of EA1 and EA2* practical work.

Students are exposed early in their first year to the idea that circuit simulation is an important aid to design, allowing design verification, evaluation of circuit sensitivities and of the consequences of faults. Initial experience is provided via an in-house analysis program (LINAC) and a commercial BBC-based version of SPICE (MITEYSPICE). In the future, once students in the

---

*As recommended by the Committee of Inquiry into the Engineering Profession, London, chaired by Finniston, H.M.S.O. Cmnd 7794 (Jan, 1980).

second year have used our SUN-based open-shop facility and the UNIX operating system, they will be able to access HSPICE via the MINNIE front-end interface (once these packages are released through the ECAD initiative).

During the 2nd-year laboratory programme, students extend their experience of simulation to include the use of HILO, the digital structural and functional simulator, and ELLA, the functional/behavioural simulation language — both of which are provided under the ECAD initiative. As part of the 2nd-year Digital System Design course, students are required to perform a four-person group digital design exercise of medium complexity. The paper design using TTL integrated circuits is submitted and assessed as a coursework assignment at the end of the Spring term but also forms the basis for a 4-week project which runs towards the end of the academic year. In the project, which forms an important part of the Department's 2nd-year EA2 programme, students have a formal opportunity to simulate parts of their system functionally using ELLA or HILO, and implement parts of designs using an ALTERA EPLD. In addition they use our in-house gate-array design teaching system which runs on BBC and Archimedes computers to lay out a part of their design on a gate array. Students have also been encouraged to design a PCB on which a discrete version of their system could be assembled.

The details of this four week laboratory based project are currently being reviewed but in the future it is anticipated that earlier access to HILO will mean that more time will be available for gate-array activities and we therefore anticipate that students will be able to use Qudos Quickchip Plus on the SUNs and on the recently installed Archimedes A440 to implement part of their design on a gate-array. Where possible a number of individual designs will be merged and sent for fabrication. After the students have finished their exams or at the start of their final year the returned chips will be tested.

3 THIRD YEAR ACTIVITIES

In the final year, students take a core of four lecture courses, totalling 80 hours, and then choose from a number of options to make a total lecture course commitment of 200 hours. VLSI work develops from the foundations of the 1st and 2nd years with a core course:

Semiconductor Devices   (20 hours)

Three further 20-hour courses are optionally available to students wishing to specialise their study and experience in IC design:

Integrated Circuit Logic Design   (20 hours)
VLSI Architectures   (20 hours)
CAD for VLSI   (20 hours)

The core course continues the two themes of Semiconductor Devices and Digital Systems beginning in the 1st and 2nd years with treatments of the process technologies involved in VLSI circuit fabrication, full- and semi-custom design styles, and the use of CAD tools to manage the design processes efficiently. Developments which have made VLSI realization accessible to even small-budget projects are described. E-Beam direct-write processing, sophisticated user-friendly CAD tools, process/technology-independent design methods, and the use of design rules are all examined in this context. Emphasis is placed on the available styles of semi-custom design and CAD tools to assist the circuit/system designer who is not required to become an expert in the details of physical device layout. The overall aim of the core course is thus to make our emerging graduate engineers aware of the options and practical trade-offs that should properly affect their thinking when faced with the implementation of systems for which they are responsible.

The three optional courses can be taken as a coherent package providing a significant degree of final-year specialization in IC design and technology, but may equally be taken individually according to the students' interests.

The first optional course, Integrated Circuit Logic Design, complements the core course in providing a treatment of silicon level NMOS and CMOS circuit design and layout. The course aims to equip students with an appreciation of a range of design techniques at the leaf-cell level, and some discussion is included of sub-function design of cells to be used in ALUs, multipliers and memories. Higher-level design considerations are addressed in the second optional course, VLSI Architectures, which considers chip and module-level design. Topics such as architecture-level design, global placement and routing, and communications strategies are used to introduce the highest-level design problems, whilst at a lower level, detailed module design and synthesis is considered. The course primarily addresses the design of digital chips, but analogue IC design is not neglected and an introduction is given to the design styles and problems associated with this class of device. The course is unconventional in that it involves the students in a programme of study which is wholly assessed by practical assignments and an open-book examination.

The final option, CAD for VLSI, deals with the algorithms and methods underlying standard CAD tools including simulators (digital and analogue); automatic placement and routing suites; and covers other CAD topics including tolerance design, sensitivity and fault analysis, test pattern generation, and symbolic methods for full-custom chip design.

All students taking these options may choose to engage in a final-year chip fabrication project, and hands-on assignment experience is organized so as to provide the required training in the appropriate tools. For the two academic years in which fabrication projects have been run using the ECAD routes to silicon, the project designs have been restricted to gate arrays. In 1987/88, four separate designs were implemented using the Silvar-Lisco SL2000 (mounted on a Microvax) and Qudos QUICKCHIP suites provided through the ECAD initiative. The Qudos system, currently mounted on an Acorn Cambridge workstation, was used in the 1986/87 academic year, and four designs were successfully fabricated by Qudos of Cambridge, using a Direct-Write E-Beam system. We have found that Qudos are able to provide a very rapid turn-round (less than two weeks for some devices in 1988). In 1986/87 all four designs were committed to TAHC06 680 gate devices, whereas in 1986/88 three designs were targetted at TAHC10 (1000 available gates) and one design at the TAHC06 device.

Timescales are defined by the timing of the project within the academic year. Projects begin in week 1 of the Autumn term (terms are 10 weeks long) and are to be concluded by week 9 of the Spring term. Students were set as a target week 10 of the Autumn term for the completion of the array designs, with a view to despatch for fabrication early in the new year.

Designs implemented over the two years have included serial and parallel multipliers, a vertical interval timebase code reader, a complex road junction signals controller and sequencer, a 5-bit fully-testable ALU, a coin totaliser to interface to a LCD display, and a hardware routing accelerator.

In all cases the students were encouraged to observe a hierarchical structure in the design process and to pay particular attention to the testability of their design. Resultant circuit complexities were of the order of 400–700 gate equivalents. Where possible the smaller designs were committed to the TAHC06 array although this has sometimes resulted in high gate utilisation and difficulties in fully aytomatic routing.

As an example, the 5-bit ALU design carried out in 1987/88 was committed both to the TAH06 and TAHC10 arrays resulting in 67% and 41% utilisation respectively, using the Silvar-Lisco GARDS route. In both cases it was necessary to revise the placement, particularly of I/O pins, in order to produce a design which was 100% automatically routed by the GARDS router. Obviously the 680 implementation required more careful placement than the 1000 gate array but in the end both were successfully achieved. Autorouting times of two to three hours were typical for the GARDS router running on an enhanced Microvax 2. Static tests on the returned Qudos produced devices indicated 2/5 of the TAHC06 and 4/5 of the TAHC10 worked successfully.

The coin totaliser was required to accept signals representing coins having a value from 1p to £1 and to produce outputs for an LCD display driver to allow the display of the current coin value and the total coin value since the last reset. Unfortunately even though the circuit was simulated using HELIX and routed on a TAHC10 device by GARDS none of the returned devices worked. This, we must emphasise, is not the result of a Qudos manufacturing problem, but is attributable to an error in circuit design and a failure in project management. The errors came about because of the use of cross-coupled inverters as latches to catch the coin signals at the input. These were driven directly by the I/O pad driver which we believe prevented the latch from operating as expected. This illustrates a problem with the HELIX simulator which does not allow for the inclusion of pad I/O drivers in the simulation. The project management failure occurred because the design review scheme adopted was insufficiently detailed to pick up the potential design problem. In future a much tighter final design assessment will be introduced.

The most complex of the designs to date is the hardware accelerator designed to be interfaced to a desk top computer to speed up the routing of gate arrays. The final design required in the order of 700 gate equivalents; unfortunately the single student on this project had great difficulty in managing the complexity of routing this design using the Quickchip system on the Acorn Cambridge Workstation and eventually had to re-enter the design using the SL2000 suite on the Microvax. He was then very short of time, and although he successfully simulated his design he was only able to achieve automatic routing leaving five unrouted connections. This ambitious project has therefore not been committed to silicon so far but is to be continued during 1988/89.

The actual fabrication of student designs has illustrated the need for adequate device testing facilities. Several designs have been fabricated that did not behave as expected. In the case of the traffic flow controller, the device testing facilities were adequate to identify a 4-bit counter that was operating incorrectly.

As noted below, we intend to move gate-array design and fabrication experience into our second year programme using Quickchip Plus on the SUN, which will then leave scope for final year project work to address topics in full-custom design. Research activity in the Department's CAD Group includes involvement in an ESPRIT project on advanced chip design environments. One by-product of this is that the Department has available the ASTRA full-custom design suite produced by British Telecom[2]. The CAD Group is involved in a number of extensions to this package to enhance its use as an inter-active *Design Assistant*-type silicon compiler[3]. Students have access to ASTRA via Apollo workstations and can experience high-level system capture, automatic high-level HDL generation and automatic floorplanning, and low-level leaf-cell design which is effected via a STICKS-based symbolic cell editor. In addition we have obtained MAGIC from the North West Universities Consortium and STIX from GEC. During the coming year, the use of these will also be investigated with fabrication being provided by European Silicon Structures (ES2) through the London University Consortium.

4 CONCLUSIONS

Experience gathered so far in operating the programme set out in Sections 2 and 3 suggests that:

(i) it is quite possible to provide a convincing VLSI Design course within the offerings of the ECAD initiative, and

(ii) we are, in general, achieving our CAD training objectives in respect of all our students (not just those specializing in VLSI Design).

Students emerge from the third year with an awareness of the crucial role of simulation tools in the design of electronic systems, and with an understanding of the trade-offs involved in deciding on a suitable style of custom design where a system is to be implemented in the form of an integrated circuit.

Students specializing in the final-year options appear to derive enjoyment from the experience that they gain in chip design. Against this, only about 10% of students currently opt for this specialization, the majority appearing to believe that they will never be involved in a real chip design. This may change as students become more aware of the rapid growth in easy-to-use design environments and silicon-compilation systems.

Our plans for the future include allowing the gata array fabrication activities to devolve into the 2nd year of the course, leaving the way clear for final-year option students to become involved in full-custom design. We envisage that this will involve students in leaf cell rather than whole chip design, but with their own cell designs added to larger devices which can then be fabricated via, for example, the route provided by European Silicon Structures. We hope to introduce this type of activity in 1988/89 using our SUN and Apollo workstations.

REFERENCES

[1]   *Standards and Routes to Registration (SARTOR)*, The Engineering Council (1984)
[2]   Revett, M. C., 'Custom CMOS design using hierarchical floorplanning and symbolic cell layout', *Proc. IEEE Int. Conf. on Computer Aided Design* (Nov., 1985)
[3]   Massara, R. E., Patel, M. V. and Nadiadi, Y., 'CAD tools for the high-level interactive design of custom ICs', *ESPRIT Deliverable Report, Dept. of Electronic Systems Engineering, University of Essex* (1987)

ABSTRACTS–ENGLISH, FRENCH, GERMAN, SPANISH

The development of an undergraduate curriculum for VLSI design

This contribution describes the way in which the ECAD facilities have been exploited in the development of an undergraduate curriculum for the teaching of IC design in an Electronic Systems B.Eng. degree scheme. The programme of course and laboratory work culminates in a final year core course in Microelectronic Circuit Technology and Design together with a specialist option in VLSI Systems Design which includes chip fabrication activities.

Le développement d'un programme de cours au niveau bachelier en conception de circuits VLSI

Cette contribution décrit la façon dont les ressources ECAD ont été exploitées dans le développement d'un programme pour l'enseignement de la conception de circuits intégrés dans un diplôme de B.Eng. en Systèmes Electroniques. Le programme des cours et travaux de laboratoire culmine en un cours central de dernière année en Conception et Technologie de circuits micro-électronique avec une option pour spécialistes en Conception de systèmes VLSI comprenant la fabrication des puces.

Die Entwicklung eines Studentenlehrplans für VLSI-Entwurf auf der Grundlage der ECAD-Initiative der Studienbeihilfe- und Staatsunterstützungsbehörden

Dieser Beitrag beschreibt die Weise, in der die ECAD-Möglichkeiten für die Entwicklung eines Studentenlehrplans zum Unterricht des integrierten Schaltungsentwurf in einem B.Eng.-Gradprojekt der Elektroniksysteme ausgenutzt wurden. Das Programm der Kursus- und Laborarbeiten kulminiert in einem letztjährigen Kernkursus der Mikroelektronik-Schaltungstechnik und Konstruktion, zusammen mit einer wahlfreien Spezialisierung auf den Entwurf von VLSI-Systemen einschliesslich Chip-Fabrikationsaktivitäten.

El desarrollo de un curriculum de pregrado para el diseño VLSI

Esta contribución describe el modo en que se han utilizado las posibilidades del ECAD en el desarrollo de un curso de pregrado para la docencia del diseño IC en un grado de B.Eng. en Sistemas Electrónicos. El programa del curso y el trabajo de laboratorio constituyen el núcleo de un curso, impartido en el último año, sobre Tecnologia y Diseño de circuitos microelectrónicos juntamente con una opción especializada en Diseño de Sistemas VLSI que incluye actividades relacionadas con la fabricación de chips.

THE INTEGRATION OF CAD WITHIN AN ELECTRONIC ENGINEERING DEGREE COURSE

G. E. TAYLOR, A. FLEMING and K. SELKE
Department of Electronic Engineering, University of Hull, England

1 INTRODUCTION

The increasing availability and use of application-specific integrated circuits has led to a demand for graduate electronic engineers with skills in VLSI design. Teaching such skills requires individual 'hands-on' experience of the various CAD tools available and this clearly presents a problem where large undergraduate classes are concerned. This paper describes the experiences of one department in tackling the problem.

The next section sets the scene with a brief description of the department and its courses. This is followed by a review of the available hardware and software and then the various areas of the course in which digital CAD plays a part are examined in more detail.

2 ELECTRONIC ENGINEERING AT HULL UNIVERSITY

The Department of Electronic Engineering at Hull University offers a four year undergraduate course leading initially to the degree of B.Eng. Students with a sufficiently good academic and practical performance have the opportunity of submitting a dissertation leading to the degree of M.Eng in a minimum period of six months from graduation. The first three years of the course are common to all honours students, with the final year split into four options: electronic communications engineering, optoelectronic engineering, electronic control and robot engineering and microelectronic and computer engineering. The Department has a commitment to the ideal of introducing all students to the use of CAD and, for example, control system design laboratories have used a modified version of the UMIST package since 1980[1] and analogue simulation has formed a significant part of design and build practicals for over a decade. The recent ECAD exercise, together with an arrangement with Plessey, has made it possible to make digital design tools available and it is with these that the current article is concerned.

3 HARDWARE AND SOFTWARE

A grant from the University Grants Committee (UGC) was used to purchase five Apollo workstations (one DN550 and four DN300). The decision to choose Apollo was governed mainly by the fact that the machines were also to be used to alleviate CAD teaching problems in our sister department of

Engineering Design and Manufacture who had already installed two Apollo workstations. Access to the machines is on a booking basis with one-hour slots available between 9 a.m. and 5 p.m. during weekdays. Priority is given to Electronic Engineering students in certain periods, coincident with laboratory scheduling, and the same courtesy is extended at other times to Engineering Design students. The package most commonly used by Electronic Engineers is HILO, although Silvar Lisco software is also available. The machines tend to be fully booked most of the time with students using these packages and hence it is currently the practice to discourage their use for program development except where this is done by postgraduates working outside normal hours. The Apollos are currently being upgraded to five DN400s and, although this will not directly alleviate the problems of machine availability, it will permit us to consider the use of a wider range of packages (including SOLO 1000/1200 and MCE's BX software) on these machines. Retention of the existing Apollos (without maintenance) provides a limited opportunity to run other software and, in the short term, ease availability problems.

In addition to the ring of Apollo workstations the department has a VAX 750 and a microVAX available for undergraduate use. Plessey have supplied their gate array design suite CLASSIC to run on these. This is part of an arrangement with a group of Northern Universities, led by Newcastle, and includes an agreement by Plessey to fabricate one or two designs per institution per year provided these are chosen from a list supplied by the firm.

Finally, SPICE is available on the University mainframe, an ICL 3980, for students wishing to examine cell behaviour in analogue detail.

4   THE SECOND YEAR DESIGN EXERCISE

Second year students are required to carry out two major 'design and build' exercises (one analogue and one digital) as a part of their normal laboratory schedule. In each case successful simulation is a prerequisite to building the circuit and thus students are introduced to:

(i)   the concept of using simulation to verify and evaluate a design.

(ii)   the specific programs SPICE (for the analogue exercise) and HILO (for the digital exercise).

Both SPICE and HILO are, of course, highly complex packages with multiple options and multiple parameter choices which can prove overwhelming to somebody meeting them for the first time. To avoid this problem students are supplied with locally produced 'teach yourself' handouts which detail, with examples, a sufficient subset of commands to complete the exercises. Copies of the full manual are supplied for those who develop an interest in CAD and wish to explore the packages further. Students work individually on the designs, each of which is allocated five weeks (30 hours) of laboratory time and the laboratory rota is such that only a quarter of the students (about eighteen people) are involved in an exercise at one time. This still causes some problems with machine time on the existing Apollos, particularly as usage peaks, naturally, in the first couple of weeks of the exercise, but the majority of students find that, with adequate preparation, 3–4 hours of actual terminal time is sufficient.

It must be emphasised that this work is intended as an introduction, only, to simulation used for design verification. The ideas of test pattern generation and verification are not introduced. A copy of the laboratory sheet for the digital design exercise, can be obtained from the authors in request.

5   THE THIRD YEAR GROUP DESIGN EXERCISE

Third year honours students have sixty hours (half a normal year's laboratory time) scheduled as a group exercise. Groups comprise five or six individuals allocated alphabetically. Half the class is involved each term and the exercise is based around the design of a digital device of sufficient complexity to ensure that the result must be the work of the whole group rather than a single individual. Management structure and task allocation is left entirely to the group and progress is assessed at a number of points during the exercise, primarily:

(i)   completion of the initial design — at this stage students are also asked to estimate required computer connect time, man hours and 'consultancy' (the amount of demonstrator or staff help required)
(ii)   when individual sub-blocks are completed and working
(iii)   when the complete device is working.

Fifty percent of the available marks are assigned on a group basis (using both reviews and final report), the remainder on an individual basis (from the final report). We are now running the exercise for the third consecutive year. Initially it was based on the CLASSIC design suite, however, greater availability of the APOLLO network means that this year HILO will be used by all students. At this stage it is intended that they extend their knowledge of the package to include such facilities as library usage and also start thinking about the problems of test. No formal teaching is given, but during the assessment reviews students are introduced to the ideas of fault simulation and encouraged to build up heuristic test sequences using their design verification patterns as a basis followed by simple ideas of path sensitisation to pick up remaining faults. A copy of the student instructions for this exercise which was discussed in more detail in Ref. [2] can be obtained from the authors on request.

Thus, all students following the honours course obtain a basic hands-on experience of using digital simulation as a design tool. Additionally, they have some understanding of the ideas of test. A much more detailed and rigorous exposure to the range of CAD tools is reserved for final year students specialising in microelectronic and computer engineering. This is a much smaller group (typically 10–20 students) and it is thus feasible to allow them extended access to a much wider variety of programs and to include fabrication of a subset of designs as part of the course.

6 FINAL YEAR CAD COURSE

As a part of the microelectronic and computer engineering option students are required to attend a course on VLSI design which is essentially a critical reivew of the available CAD tools. Topics covered include simulation (true value and fault), automatic test pattern generation, design for testability and self test and the course text is *CAD for VLSI* by Russell et el[3]. The course is supplemented by:

(i)   open access to the various packages provided by the ECAD initiative with exercises involving modelling for simulation, test pattern generation and verification etc.

(ii)   an assessed design exercise in which students are asked to design a medium scale device (typically a counter or register) with attention paid to timing, testability, generation of test sequences etc. as well as to functionality.

The mark from the assessed exercise, which currently runs using the CLASSIC suite, counts as the equivalent of one question in the final examination (i.e. approximately 30% of the course mark). Additional motivation is provided by choosing the two best designs to go to Plessey for fabrication.

7 M.Sc. COURSE

The Department of Computer Science at Hull has recently instigated an M.Sc. course intended as a conversion course for non-computer science graduates. One module of this course is concerned with the computer aided design of digital circuits. After an introduction to digital circuit theory the syllabus covered is essentially that of the final year course described above, extended to allow for inclusion of the additional background material. Again, open access to the various programs is given and students are encouraged to experiment with the different algorithms available and to find their limitations.

8 CONCLUSIONS

The Department now has several years' experience in introducing students to CAD tools in the way described. By making initial teaching and access part of the laboratory schedule every student learns something of the available tools and yet the worst problems of machine overload are avoided. At this stage, discrete rather than integrated circuits are considered, but the lessons learnt are equally applicable to ICs. The in-depth exposure to the full range of packages is reserved for the single final year option for whom it is deemed most appropriate and this again means the number of students is manageable. The group is currently exposed to all parts of the IC design and production process except the use and problems of automatic test equipment. This is something we hope to address as more time and funding becomes available.

REFERENCES

[1]   Taylor, P. M. & Taylor, G. E., 'The use of a computer aided control system design suite in an electronic engineering degree course', *Proc. 4th IASTED Symposium on Modelling, Identification and Control, Grindelwald, Switzerland* (1985)

[2]   Taylor, G. E., Fleming, A. & Mulvaney, D. J., 'A gate array design laboratory', *Proc 5th Conf. on The Teaching of Electronic Engineering in Degree Courses — Shaping for the Future*, University of Hull (1988)

[3]   Russell, G., Kinniment, D. J., Chester, E. G. & McLauchlan, M. R., *CAD for VLSI*, Van Nostrand Reinhold (1985)

ABSTRACTS–ENGLISH, FRENCH, GERMAN, SPANISH

The integration of CAD within an Electronic Engineering Degree Course

The Electronic Engineering Department at Hull offers a course which is committed to introducing students to concepts of CAD from early in their undergraduate careers. The paper describes ways in which various CAD tools, particularly those made available via the ECAD initiative, are integrated into different stages of this course.

Intégration de la CAO dans un cours du diplôme d'ingénieur électronicien

Le Département d'électronique à Hull offre un cours qui a pour but d'introduire les étudiants aux concepts de la CAO dès le début de leur carrière estudiantine. L'article décrit la manière dont les différents outils de CAO, en particulier ceux mis à disposition par l'initiative ECAD, sont intégrés aux différentes étapes de ce cours.

Die Integration von CAD innerhalb eines Elektronikingenieur-Graduierungskurses

Das Department für Elektronikingenieure von Hull bietet einen Kurs an, der Studenten bereits frühzeitig vor der Graduierung in CAD-Konzepte einführt. Der Beitrag beschreibt Wege, auf denen verschiedene CAD-Tools, vor allem solche, die über die ECAD-Initiative verfügbar werden, in verschiedene Stufen des Kurses integriert wurden.

La integración CAD en un curso de Ingenieria electrónica

El departamento de ingenieria electrónica del Hull ofrece un curso que introduce a los estudiantes en los conceptos del CAD desde sus principios en los primeros cursos. El articulo describe las formas por las cuales diversas herramientas de CAD, particularmente aquellas disponibles mediante la iniciativa ECAD, se integran en los diferentes apartados de este curso.

THE INTRODUCTION OF ELECTRONICS COMPUTER AIDED DESIGN FACILITIES TO ENGINEERING SCIENCE STUDENTS

ROSEMARY A. COBLEY
Department of Engineering Science, University of Exeter, England

1 BACKGROUND

The broader nature of an Engineering Science degree provides a different framework and structure for education and training to that embodied in a traditional Electrical and Electronic Engineering undergraduate course. The department of Engineering Science at Exeter University currently runs both theoretical and practical courses related to electronic engineering. All students of the degree course take a common first two years of subjects, covering aspects of electrical, mechanical and civil engineering science. During the final two years of the course more specialised options can be taken plus other topics including engineering management. Currently there are courses in the core stream related to electronic theory and electronic design[1]. As part of the department's computer aided engineering (CAE) programme[2], new final year courses on computer aided design (CAD) and large scale integrated (LSI) circuit design commenced in 1984 and 1985 respectively. In-house and professional electronics computer aided design (ECAD) tools have significantly enhanced the facilities available to the second year undergraduate course on Electronic Design and the final year CAE options.

This paper outlines the process of introducing the new facilities to the CAE programme. In the first section, the ECAD tools are described (both hardware and software) together with the arrangements for training. The next section assesses the impact of the tools in terms of how the courses have been adapted to take advantage of the new facilities and the new opportunities for design and manufacture in student projects. The final section takes a look ahead at the future use of the ECAD tools in the context of changes currently taking place at Exeter University, particularly the scope for testing completed projects.

2 THE ECAD TOOLS

Under the ECAD initiative, started in 1985, funds were made available for the purchase of engineering workstations by universities, polytechnics and colleges of higher education. The licences for a range of professional electronic CAD programmes were also purchased. The engineering communities were allowed access to these programmes. At Exeter University the grant from the scheme, together with departmental funds, enabled the purchase of a network of 5 Apollo DN3000 workstations. Two of the workstations had colour monitors, as it was felt that these would be essential for the more graphically complex tasks of manual routing of printed circuit board and integrated circuit artworks. Although a wide range of software was made available, the option for programmes from Silvar Lisco, Micro Circuit Engineering, Genrad and Racal was procured. This was due to an initial restriction whereby only certain programme suites would operate on the chosen workstations.

The suite of programmes chosen each offered the opportunity to explore different aspects of the design and manufacture process. The Silvar Lisco programme suite provided a comprehensive range of programmes for the schematic input of circuit diagrams, simulation of digital and analogue circuits, functional simulation of logic circuits and layout of semicustom and full custom integrated circuits. The programmes from Micro Circuit Engineering were tailored towards the schematic input, simulation and automated layout of the gate array range, manufactured through the company's silicon brokerage facility. Software for logic circuit simulation and fault simulation of circuits was available from Genrad. Finally the Racal Redac suite allowed the schematic capture and layout of printed circuit board artworks.

Familiarisation by the teaching staff of the new facilities was necessary prior to incorporation into the syllabus and their use by students. Although the discussion of the ECAD initiative commenced in 1985, the first range of software was received by Exeter University in the summer of 1986. During this period training courses were arranged for academic staff and computer support personnel by the respective software houses. System support was provided by the Engineering Science Department through two academic members of the teaching staff with occasional technician assistance.

3   THE IMPACT OF THE ECAD TOOLS

The immediate task of familiarisation of the new facilities was smoothly accomplished in the case of the schematic input tools and the electronic simulators. This was assisted by example circuit databases, upon which the tools could be tested, being provided by some of the software suites. The ECAD tools were therefore introduced during the 1986/87 session in the optional courses in Computer Aided Design and LSI Circuit Design.

At the outset, the method for the incorporation of the new facilities into the existing courses was assessed. To accommodate this, the module on 'Electronic CAD' was substantially revised[3]. This module discussed the techniques for the storage of electronic circuit data and the design and manufacturing techniques applied in industry for the production of electronic equipment using printed circuit boards. As the students were now dealing with complex software suites, there was a need to provide a series of 'getting started' guides for their use. This was in addition to the introduction to the use of the ECAD facilities.

Previously, the practical aspects of ECAD in the CAD course were primarily based on the use of videos and visits to local electronic companies, who were using ECAD tools. The theoretical aspects of the module included the discussion of the computer based techniques for circuit design and circuit manufacture used in industry. Although the visits proved popular, the students had little opportunity for 'hands-on' experience of the tools.

However, with the introduction of professional ECAD tools, the students could now tackle small projects involving the design and simulation of a digital circuit to a given specification. An example circuit project is shown in Fig. 1, where the circuit is designed to provide a four bit binary to Gray code conversion. The projects were programmed over a period of two weeks to allow access to the workstations by the course students. The final output from the project was in the form of a circuit diagram and a copy of the simulator listing which indicated the required input waveforms and checked the circuit outputs for the required response. The projects allowed a more rigorous involvement in the design process and reinforced their understanding of the designed circuit operation.

In the LSI Circuit Design course, prior to 1986, the students studied the theory underlying the design of LSI circuits. This included discussions on the practical techniques used in industry for LSI design. The course project involved the use of a two dimensional draughting system, which had been developed in-house, for the schematic input of circuits. This was interfaced to a simplified logic simulator. The final chip was laid out using the draughting system, with a gate array layout plot as background, onto which metallisation tracks were drawn.

The introduction of the ECAD tools resulted in the students being assigned a group project for the design of a digital system to perform a given specification. The digital system was split into discrete circuits for which individual students were responsible. Example systems have included a railway signalling system[4] and a telephone switching network simulator. An example design from the

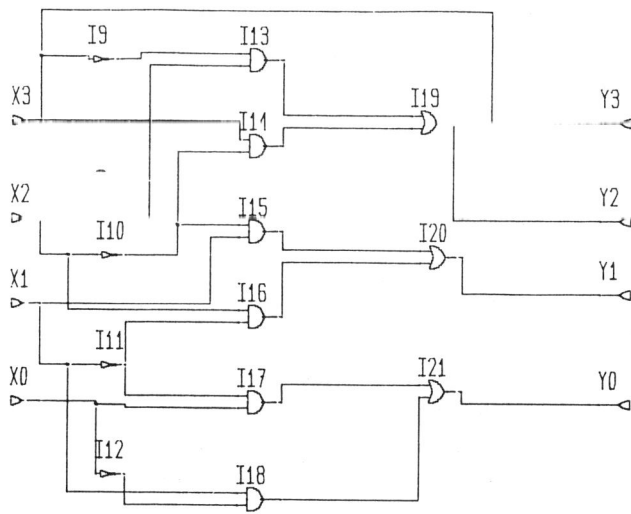

FIG. 1   A binary to Gray code circuit from the CAD course

signal control section of the former project is shown in Fig. 2. The opportunity now existed to take the design process a stage further and besides being able to simulate the results, through the use of the newly acquired software, to actually build the gate array. Co-ordination of the separate designs was necessary to ensure that the deadline for the submission of the designs to the silicon broker-age was met. Particular importance was attached to the correctness of the design as demonstrated by the simulator data. With lead times of four weeks for manufacture, the testing of the devices was completed after the student examinations.

The CAD course students' experience of the ECAD tools resulted in a degree of investigation of circuit characteristics without the necessity of producing a 'breadboard' circuit. Their impressions of the tools were so favourable that several wished to use them for the completion of final year project designs. In the case of the LSI design course students, the access to ECAD tools enabled the design of more complex systems than would normally be possible during the course's duration. The tools allowed the students to go further in closing the design loop.

5   FUTURE USE OF ECAD TOOLS

The introduction of the ECAD tools into the CAE programme has provided the opportunity for their further use in other areas. The previous sections have covered the initial use of the ECAD tools. However this is an evolutionary

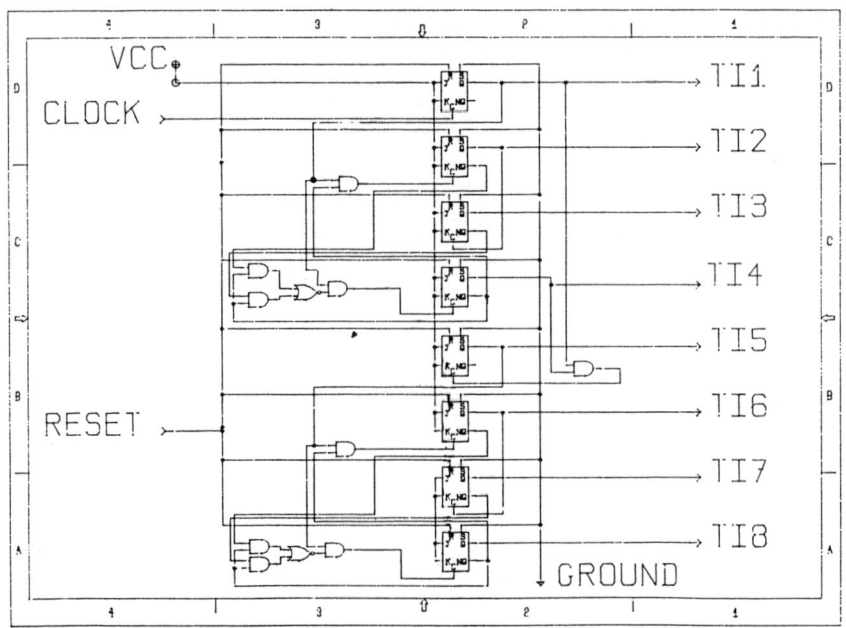

FIG. 2   An 8-bit BCD timer circuit from the LSI course

process and a number of changes are currently taking place. These relate to increased software availability and new releases together with developments occurring within Exeter University.

Under recent developments, software from European Silicon Semiconductors (ES2), Praxis and Meta-Software will now operate on Apollo workstations. This enables the introduction of full custom integrated circuit design at a reasonable cost for undergraduate research work. In addition to the digital circuit design software from ES2 and Praxis, a more user-friendly analogue circuit software (HSPICE) from Meta-Software will also enable more accurate analogue circuit analysis to be performed. A more general benefit however is the considerable interest shown throughout the department by academic/technical staff and research students. Consequently, members of the department have been provided with their own programme of introductory courses. These have focussed on the areas in which the ECAD tools can be applied given the specific interests of the department. The interdisciplinary nature of the department has created new possibilities for student projects and research initiatives. These have often emphasised the linking of ECAD tools to related CAE areas.

As in the case of software availability, the teaching requirements at Exeter University are undergoing important changes. During 1988 the Engineering Science and Chemical Engineering departments will be combined into a newly created School of Engineering. Consequently, the existing courses will be restructured. The initial two years of the degree will remain a core course with optional courses in the final two years. Engineering design activities will be introduced throughout the two years of the core course.

The implications of these changes are that the students will have access at an earlier stage to ECAD tools. This is planned to take the form of laboratory 'design, simulate and verify' experiments for student groups. During the core course there will be a greater emphasis on computer aided instruction in the use of the ECAD tools. In the optional courses of the final two years there is the possibility to introduce the more recently made available software, alongside that already used, within the CAD and LSI modules. Also it is intended to extend the design process into the teaching of circuit testability by the continued use of in-house software supplemented by appropriate professional software, should that become available. However, interfacing of the tools for circuit testing and the provision and access to adequate test equipment are still unresolved. The interest and involvement of the Industrial Advisory Committee in the ECAD initiative will provide a source of new areas for student projects that will be of relevance to the needs of industry and the wider interests of future graduates of the new School of Engineering.

6 CONCLUSIONS
The introduction of the ECAD tools has enhanced the teaching capability in areas of Electronic Design and LSI Circuit Design at Exeter University. Existing CAE courses have been adapted to incorporate the new facilities and new opportunities have been created for student project work.

The impact of the ECAD tools has been considerable in the relatively short time period that the facilities have been available to the department. Integrated circuits have been built and some preliminary testing work undertaken. However these achievements must be set within the continued need for new developments and against a background of further change.

At present, two sets of changes are taking place. The first concerns the availability of new software and the need to introduce this into the course structures alongside that already in educational use. This involves incorporating the ECAD tools in the design and simulation of combined analogue/digital devices. The second relates to changes in course structures resulting from the new syllabus devised for the School of Engineering. Further enhancement of the CAE programme is envisaged involving tackling problems relating to the interfacing of the ECAD tools to circuit testing and the use of test equipment.

The ECAD initiative has contributed toward the process of updating and review of departmental facilities required for the education and training of undergraduate students in Engineering Science. The continued introduction of appropriate tools will greatly benefit those emerging from the new School of Engineering into the demanding world of modern engineering.

7 ACKNOWLEDGEMENTS

The progress made on introducing ECAD facilities at Exeter University has been made possible by the help and support of a wide range of colleagues in the Department and elsewhere. I would particularly like to acknowledge the encouragement given by Prof. D. J. Woollons.

8 REFERENCES

[1] Bland, R. J. and Cobley, R. A., 'The introduction of electronic design to an engineering science core course', *IJEEE*, **22**, pp. 293–298 (1985)

[2] Cobley, R. A., 'The development of a Computer Aided Design course for undergraduate students of Engineering Science', *CAE Journal*, **2**, No. 4, pp. 118–121 (1985)

[3] Cobley, R. A., 'The integration of professional software into a CAD Course for Engineering Science students', *CTISS Workshop on Computers in Engineering Education, Imperial College, London* (1987).

[4] Cobley, R. A., 'The application of fault testing to a railway signalling system', *IEE Colloquium on 'Design for Testability'*, Digest No. 1988/32 (March 1988)

ABSTRACTS–ENGLISH, FRENCH, GERMAN, SPANISH

The introduction of ECAD facilities to Engineering Science students
The impact made by professional electronic CAD tools and engineering workstations, upon undergraduate courses in the department, is outlined. Student groups have successfully designed, simulated and tested digital systems, which had been implemented as gate arrays. The interest throughout the department has highlighted areas for integrated mechanical/electronic projects.

Introduction des moyens ECAD à l'usage des étudiants en Sciences Appliquées
L'impact obtenu par les outils professionnels de CAO en électronique et les stations de travail en ingéniérie, sur les enseignements du département est esquissé dans cet article. Des groupes

d'étudiants ont, avec succès, conçu, simulé et essayé des systèmes digitaux, qui avaient été réalisés comme réseaux prédiffusés. L'intérêt éveillé dans le département a mis en lumière des possibilités de projets intégrés mécanique/électronique.

Die Einführung von ECAD-Einrichtungen für Studenten der Ingenieurwissenschaften
Der Einfluß von professionellen electronischen CAD-Tools und Workstations auf die Ausbildung in Kursen noch nicht graduierter Studenten im Department wird ausgewiesen. Studentengruppen haben erfolgreich digitale Systeme entworfen, simuliert und getestet, die als Gate-Arrays implementiert wurden. Das durch das gesamte Department hindurchgehende Interesse hat Gebiete für integrierte mechanisch-elektronische Projekte herausgehoben.

La introducción de las facilidades ECAD a estudiantes de ciencias de la ingenieria
Se explican el impacto de la herramientas ECAD profesionales y las estaciones de trabajo Ingenieriles en los cursos del departamento. Grupos de estudiantes han diseñado un simulador y comprobado con éxito sistemas digitales que han implementado con gate-arrays. El interés general del departamento ha remarcado áreas para proyectos de Meccánica/Electrónica integrada.

ASIC DESIGN, THE SOLO 1000 TOOL SET, AND BTEC COURSES

G. L. LAWDAY* and R. G. FORBES†
*Bracknell College, Bracknell, Berkshire, England
†Department of Electronic and Electrical Engineering, University of Surrey, England

1 INTRODUCTION

1.1 General background

Application Specific Integrated Circuits (ASICs) are increasingly used as components in electronic products of many kinds. The basic reasons for this need no detailed rehearsal here, but include reduced component costs, enhanced product reliability, and often reduced product size. Alternatively, ASICs may contribute to getting a more sophisticated product for a given product cost. Time to market can also be reduced.

Part of the reason for the growing popularity of ASICs lies in the increasing sophistication of the software support available. Apart from speeding up the whole ASIC design process, modern design packages try to make the design process more 'accessible'. The hope is that designing ASICs will become less of a specialist occupation, and more of a skill than the 'ordinary' design engineer can acquire without undue expenditure of effort.

These arguments apply with rather more force to the 'semi-custom' ASIC design techniques, of which gate-array and standard cell techniques are the best known examples, than they do to the production of fully handcrafted chips. Our concern here is with semi-custom techniques, and in particular with the 'optimised array' technique developed by European Silicon Structures (ES2). The Higher Education ECAD (Electronics Computer Aided Design) iniative was a recognition that we now need to provide information about ASIC design techniques to undergraduates who intend to become chartered electrical engineers, and to give them practical experience of such techniques.

In the workplace, the graduate or chartered engineer will often be backed up by supporting personnel, of incorporated engineer or engineering technician level, and there is a corresponding need to provide information about ASIC design methodology in their education and training. This need is evident from national surveys[1,2], and is also evident from local employer demands in Bracknell, where one of the authors is a lecturer at Bracknell College.

There is also a need for graduate post-experience courses, both for updating the experienced engineer in modern design methodologies, and for continuing the training of the freshly employed graduate, along the lines of the old 'EP1' courses in the U.K. These are now called 'AET' — Applications of Engineering Technology — courses.

Stimulated by this local employer demand, one of the authors has, for the last two years, had responsibility for introducing courses in semi-custom design methods into the teaching at Bracknell College. In the coming (1988/89) session, the College will offer two new BTEC Higher Certificate* final-year modules — one a teaching unit in semi-custom design, the other a semi-custom design project — as part of a new Higher Certificate course in 'Engineering (Software)'. A new 90hr BTEC 'Continuing Education' course in the semi-custom design will also be offered.

The relevance of these developments in the context of this special IJEEE issue is as follows. These are the first BTEC-approved modules in semi-custom design to be offered in the U.K.; they are being offered in the Further Education sector rather than the Higher Education sector; and practical design work will be based on the SOLO 1200 toolset marketed by ES2. This toolset has recently become part of the Higher Education ECAD initiative, and Surrey University will be one of the lead sites.

The intention of this article is to describe the background to these developments, and our experiences so far. We begin with some comments on the BTEC system, to set the context for those unfamiliar with BTEC courses in the U.K.

1.2 BTEC courses

BTEC courses have a modular structure. As from September 1988, modules are classified as N-level or H-level. The former are primarily intended for National Certificate (or Diploma) courses. The latter are intended for Higher Certificate (or Diploma) students, who come in from National Certificate or A-level studies. BTEC National and Higher Certificate modules are quantified in terms of a notional 60-contact-hour unit, individual modules counting between 1/2 and 2 units. Award of a Higher Certificate requires passes equivalent to a minimum of 10 units, of which at least 8 must be at H-level. There has to be a degree of coherence about the modules taken, and the Certificate is awarded in a stated area, e.g. 'Software', or 'Electronics'. Certain modules may be designed as 'core' (i.e. essential) for the award of certain Certificates, and it is possible to specify pre-requisites and co-requisites for the study of any particular module.

A college that wishes to introduce a new BTEC Certificate course or to introduce fundamentally new modules into the BTEC system has to make a formal submission to BTEC for approval. With fundamentally new modules, BTEC sets up a special validating panel. Its members scrutinise the submission, and inspect the college requesting approval, and then (may) give permission for a two-year pilot period, during which course progress is closely monitored. If the pilot period proves successful, then the module syllabuses become part of

---

*The Business and Technician Education Council (BTEC) Diploma and Certificate engineering courses in the U.K. are designed for the education and training of incorporated engineers and engineering technicians and are less academic and more practical than degree courses.

the BTEC portfolio and are available to be taught by any college that can demonstrate that it has the necessary resources.

The two Bracknell semi-custom design modules have received approval and are about to start their pilot period. Each module rates as 1 unit. They form part of the final year of an Engineering (Software) Certificate because this is how things have developed locally, in response to employer demand. The remainder of the course consists of digital electronics, computer systems, and software topics, all units being at H-level. We consider that this mixed software/hardware context is admirably suited to a semi-custom ASIC design course, though it should, of course, be perfectly feasible to mount such a design course in a mainstream electronics context.

1.3   Staff training

Colleges, like universities and polytechnics, have difficulties in keeping their staff up to date with modern developments in high technology. Part-time attendance at a local M.Sc. course is a well-established method, though the real value of this, possibly, lies as much in the time off given by the college for private study and reading as it does in the attendance at lectures. G. L. Lawday has recently attended the Surrey M.Sc. course in Microelectronics and Computer Engineering, which includes material on MOS VLSI design and CAD. The associated M.Sc. project (which has R. G. Forbes as the university supervisor) has been slightly unusual in that it mixed engineering and education. Part of the project has been a substantial design (a counter-timer meter, of complexity equivalent to approximately 1800 gates), using the ES2 SOLO 1000 toolset; but an equally significant part of the project has been the preparation of the BTEC course submissions mentioned above, and the construction of a set of 'laboratory exercises in semi-custom design' to support the lectures.

This project conformed with the philosophy for part-time students, that projects should, where appropriate, have relevance to the needs of the student's employer. It was also envisaged that this work would have relevance to the introduction of the ES2 tools into degree-level teaching, though some difference of approach might be needed.

ES2 is situated in Bracknell, close to Bracknell College, and the design part of the project was carried out at their premises and using their equipment. As a preliminary, both of us attended a week's training course on the SOLO tools. We are grateful to Paul Gibson, the ES2 Director of Education, and to Paul Naish of the ES2 training staff, for provision of facilities and for their personal willingness to share with us their experience of electronic design and their detailed knowledge of the SOLO toolset.

2   OPTIMISED ARRAYS AND THE SOLO TOOLSET

As already indicated, the project used the SOLO 1000 toolset, which was the original member of the SOLO 1000 family. These toolsets support the ES2 'optimised array' methodology. This is a design style which feels to the user rather like a gate-array technique, but in which all mask layers are, in fact, fabricated, now normally by the use of e-beam direct-write-on-wafer techniques. It aims at the market sector where designs are of moderate (or less) complexity, and the requirement is either for low-volume production, or for rapid prototyping. It is also marketed as a design style that is easily accessible to first-time users of ASICs.

The methodology is based on the 'stage', which is a CMOS transistor pair of standardised size. Adjacent stages may be interconnected to form gate-level logic functions, and individual stages may be suppressed to provide space for a local interconnection. More complex logic functions can be constructed from the gate-level functions and stored as library items.

The stages are 'strip-farmed' into rectangular blocks, containing perhaps 100 stages, and the blocks are arranged in rows (as many as appropriate) and columns (usually 2 or 3), with appropriately adjusted routing channels between them. This central design area is surrounded by a periphery of pads.

Although facilities do exist for manual intervention (to constrain or trim up layout), placement and routing (including pad placement) is basically automatic, being carried out by the silicon compilation tools originally developed by Lattice Logic (now part of ES2).

Thus, for the SOLO user, the design process consists of the following main steps: (i) Prepare a logically correct design; (ii) Capture it, using either the schematic editor ('DRAFT'), or the system's hardware description language ('MODEL'); (iii) Prepare simulation vectors and/or waveforms, and exercise the circuit using the system simulator ('EXERT'); (iv) Iterate previous steps as necessary; (v) Request automatic layout. A set of test vectors needs to be produced, for application to the fabricated chip. There is also a software 'design process manager', which checks that all appropriate stages have been completed before the tape is finally shipped to ES2. And, obviously, there are facilities for outputting appropriate documentation.

The system expects design to be carried out hierarchically. Thus, for example, the schematic editor has facilities for the automatic creation of icons to represent a circuit module described by a schematic diagram, so that this icon can represent the module in the next higher level of the hierarchy.

The SOLO 1200 tools, which are part of the ECAD initiative, have the facility for putting down certain standard analogue cells and fixed RAM blocks adjacent to the pads. The SOLO 1400 tools will allow designs to include RAM, ROM and PLA blocks of size adjustable by the designer.

3 EXPERIENCES WITH THE TOOLS

3.1 *Initial experience*
Our initial experience with the tools used small exemplar designs. We found the tools relatively user-friendly, and with something like a 4-to-1 decoder it was possible to start from a position of no experience and get through all the stages in less than a day.

In the schematic editor, connections between circuit symbols are 'rubber-banded' (i.e. the connections remain in place as the symbols move around the screen); we found this facility particularly useful.

Our general impression was that the tools would be easy to employ in design exercises both in degree-level and higher certificate courses, and are suitable for project work.

The ES2 documentation we would rate as 'relatively good'. It has noticeably improved over the last eighteen months, and the process continues. A useful feature is that a number of 'walk-through' design exercises are now provided. The tools have been used in a Surrey undergraduate project in the 1987/88 session, and the student concerned had no difficulty using the documentation on a self-tutorial basis.

3.2 *The project*

As already indicated, the project was the design of a counter-timer meter. This was envisaged as an industrial device useful in the context of motor maintenance that would record the number of revolutions of a motor and/or its on-time, and would display the result in a number of alternative modes, as requested by the user. At the time of writing, the project has been taken as far as the layout stage, but not yet to fabrication. In terms of its digital electronics, the main project components are counters, multiplexers, buses and combinational logic.

Carrying out the project was a highly useful educational experience, for it became clear that the first step in the above process: 'Prepare a logically correct design' could usefully be subtitled as 'Prepare a correct design that will be accepted by the simulator, and is properly testable'.

It became apparent that design skills, practices and short-cuts learnt in the context of discrete logic were not necessarily transferable to the design of integrated circuits, particularly where matters of timing are concerned. A typical manifestation of this was that failure to give adequate attention to clock distribution resulted in a circuit that put the simulator into an unknown state, from which it would not emerge. The circuit in question could be presumed to work adequately when built from discrete logic, though in special circumstances it might miss a 'count'.

More generally, the tools demand a disciplined attitude because they are bureaucratic, and force the designer towards systematic hierarchical design methods.

Before the project began, we had a simplistic idea that the primary role of the ASIC design course would be to teach students how to transfer existing logic designs, largely unaltered, onto a single chip by using semi-custom design tools. However, the questions concerning timing and the need for testability were found to dictate a knowledge of hazard-free design and design for test, even for relatively simple tasks. And it became clear that considerable skill was involved in designing testable synchronous logic. The experience of the project influenced both the content of the design exercises, and the decision to include 30hrs of material on 'Design for Test' in the Continuing Education programme.

ES2 have also found that some customers need training, not only in the use of the tools, but also in more basic matters concerning sequential logic design and testability.

4 THE BTEC CERTIFICATE MODULES

As compared with the teaching of undergraduates, Higher Certificate teaching needs to concentrate in more detail on the methodology of design. Also, familiarity with the gate-array type of design methodology is much more relevant than familiarity with full-custom techniques. As indicated, a BTEC module on semi-custom design has been prepared, with the following principal objectives, namely that the expected learning outcome is that the student will:

(i) Apply a knowledge of bipolar and MOS technologies to appraise their suitability for semi-custom integrated circuits.
(ii) Describe silicon fabrication technology with particular reference to semi-custom device production.
(iii) Review the specification, design methodology, and development cycle for a semi-custom integrated circuit.
(iv) Construct circuit diagrams with a hierarchical design method, using a schematic editor.
(v) Construct circuit parts using a hardware description language and its associated compiler.
(vi) Interpret compiled HDL or schematic source code to determine network numbers and fan-out bonding.
(vii) Apply a test simulator using interactive and batch-driven modes, and then interpret its output.
(viii) Construct a silicon layout and relate the artwork to its circuit function.
(ix) Develop a critical appraisal of semi-custom integrated circuits and their applications.

In this module, the design exercises play a central role. Students work in pairs, using the SOLO 1200 tools on a Vaxstation 2000 workstation, with a contact group of eight students at a time. It was perceived that these exercises should each involve an explicit, well-documented, walk-through of some feature of the tools or of ASIC design. Also that the lecturer needs to orchestrate the process, by first demonstrating the feature in question, then keeping a close watch over student practice with the tools, and finally drawing conclusions.

A broadly similar pattern is, in fact, used with the student laboratory exercise in full-custom design in the 2nd year of the Surrey University course; it has been found that an active and much closer watch needs to be kept over how the students are interacting with the graphics screen than is normal with laboratory experiments. Eight inexperienced students are enough for one demonstrator.

The SOLO design exercises cover the following topics:

Exercise 1:   The schematic editor;
Exercise 2:   The hardware description language;

Exercise 3: Simulation;
Exercise 4: The physical chip layout;
Exercise 5: Combinational circuit design;
Exercise 6: Hazard-free design;
Exercise 7: Employing library parts;
Exercise 8: Avoiding essential hazards;
Exercise 9: Pad cells;
Exercise 10: Testable silicon designs.

Further details both of the exercise and of the module as a whole can be obtained from G. L. Lawday.

The second semi-custom design module is a 60hr project, using the SOLO 1200 tools, which is carried out in the final stages of the course. This is a vital part of the course, for it is at this stage that some of the real-world difficulties in designing semi-custom chips will be brought home to the student.

5   THE BTEC CONTINUING EDUCATION UNITS

These comprise a 60-hr course on ASIC Design, with syllabuses similar to that outlined above, and a 30-hr course on Design for Test. Both units have received BTEC approval. At Bracknell they will be combined to form an evening course for updating local engineers; the material will also be used in AET courses for new graduates. Further details can be obtained from G. L. Lawday.

6   SOME FINAL COMMENTS

As the courses described above are only just starting, it is obviously premature to think about conclusions, though it perhaps deserves reporting that enrolment on the Certificate course this year is more than double what it was last year. However, we do have some thoughts on the preparation process.

First, it has been useful to have a relatively long time to prepare for the introduction of this type of course, to allow its organiser to become familiar with the problems involved in ASIC design.

Second, the carrying out of a project by the course organiser before finishing the design of the courses has proved invaluable, for — as stressed above — it suggested that the basic teaching problem would not be familiarisation with the tools, but the need to inculcate basic principles of synchronous logic design and design for test.

Clearly, it will be advisable to set projects where the students have to grapple, if only in a small way, with problems of this type. We expect this to be true of work in the university sector too.

Two closing points — first, we have enjoyed the work described here, and we also expect most students to enjoy semi-custom design projects.

Finally, we believe that there is an important role for support staff in industry in the context of ASIC design; course provision as described here ought to become more widespread in the Further Education sector, and may also be useful in the context of Higher Certificate/Diploma work in the polytechnic sector in the U.K.

REFERENCES

[1]   Senker, P., *Employment and Training of Technicians, pt II*, Report produced by Science Policy Research Unit, Sussex University (1985)

[2]   D.T.I., *The Impact of Microelectronics*, Report produced by the Policy Studies Unit, on behalf of the Department of Trade and Industry (1986)

ABSTRACTS–ENGLISH, FRENCH, GERMAN, SPANISH

ASIC design, the SOLO 1000 toolset and BTEC courses

Bracknell College now offers the first BTEC-approved Higher Certificate modules in semi-custom design methods in the UK. Practical design exercises involve the ES2 SOLO 1000 toolset. This article outlines the content and philosophy of the modules and the preparation for introducing them, including our own experiences with the tools.

Conception ASIC, les outils SOLO 1000 et les cours BTEC

Le Bracknell College offre maintenant les premiers modules d'un Certificat Supérieur approuvés par BTEC dans les méthodes de conception de circuits pré-diffusés. Les exercices pratiques de conception impliquent l'utilisation des outils ES2 SOLO 1000. Cet article esquisse le contenu et la philosophie des modules et la préparation de leur introduction, y compris notre propre expérience avec ces outils.

Entwurf von anwendungsspezifischen integrierten Schaltungen (ASIC), der SOLO-1000-Rüstzeugsatz und BTEC-Kurse

Bracknell College bietet jetzt die ersten BTEC-anerkannten Höher-Zertifikat-Bausteine für Entwurfsmethoden von halbverdrahteten integrierten Schaltungen im Vereinigten Königreich. Praktische Entwurfsübungen umfassen den ES2-SOLO-1000-Rüstzeugsatz. Diese Arbeit skizziert den Inhalt und die Philosophie der Bausteine und die Vorbereitungen zu ihrer Einführung, einschliesslich unsere eigenen Erfahrungen mit dem Rüstzeug.

Diseño ASIC, el SOLO 1000 y los cursos BTEC

En el Bracknell College se ofrecen actualmente los primeros Certificados Superiores, BTEC, sobre diseño semi-custom en el Reino Unido. Los ejercicios prácticos de diseño utilizan el ES2 SOLO 1000. En esta articulo se perfilan los contenidos y la filosofia de los módulos y la preparación necesaria para manejarlos, de acuerdo con la experiencia adquirida con estas herramientas.

INTRODUCING ELECTRONIC CAD TO FIRST YEAR STUDENTS

L. T. WALCZOWSKI

Electronic Engineering Laboratories, University of Kent at Canterbury, England

1   INTRODUCTION

In early 1986, an Electronics Computer Aided Design (ECAD) centre was established in the Electronic Engineering Laboratories at the University of Kent, part financed by the University Grants Committee (UGC) ECAD Initiative and partly sponsored by industry. The ability to run as many of the tools as possible obtained under the Initiative had governed the choice of workstation. Eventually the decision was made to base the centre on Apollo workstations. The laboratory now consists of ten workstations, most of which are colour, three printers, an A0 8-pen plotter and over one gigabyte of disk space. The ECAD software includes Silvar Lisco's SL2000 suite and GenRad's HILO-3. Concurrently with the establishment of the new laboratory, the development of a specialised, introductory course in Electronic CAD was commenced. This course, consisting of a one day workshop, backed up by a five hour lecture course has now been running for two years, and is scheduled in the Lent term for all first year computer science, computer systems engineering and electronics students.

2   AIMS AND OBJECTIVES

From the outset, the decision was taken that the correct place for an introductory course in Electronic CAD was in the first year rather than in the two subsequent years. The philosophy behind this was two-fold: firstly, the use of CAD tools in 'real' design work is obligatory in the same sense as the use of voltmeters and oscilloscopes is in testing assembled circuits. Therefore the use of software tools should be introduced at the same time as hardware tools. Secondly, design exercises using sophisticated design tools running on powerful workstations should motivate and encourage new undergraduates, perhaps rather more than pen and paper exercises.

The major concern initially was not so much course content but the problem of resource scheduling. Originally only four workstations were available for a total of about 140 students. The workshop was thus incorporated into the first year laboratory schedule which ran on two days of the week with students working in pairs. Thus each student-pair has obtained at least five hours workstation experience by the end of the first year.

The lecture course was developed not only to give training in the use of the facilities available in the ECAD laboratory, but also to introduce in a simple way the concept of simulation at various 'levels of abstraction', and hence hierarchical design methodology albeit at an elementary level.

The examples chosen for the workshop were designed to exploit the special properties of a workstation, namely graphics and processing speed. However, since students would have no knowledge of the native operating system running on the workstations, tutorials were devised to be largely self-contained with almost all work carried out in the local environment of the design system. (The experiment, however, still requires some use of the screen editor). Since the natural starting point for any design exercise, whatever the implementation, printed circuit board, semi-custom or full custom, is entering the circuit, the corner stone for the workshop would have to be a schematic capture system. Thus, a design exercise based on circuit entry, simulation and graphical post-processing of the results was developed.

3 LECTURE CONTENT

Although the majority of the course material concentrates on several of the tools obtained under the Initiative and their role in the design of an electronic system, the first lecture is an introduction to the evolution of Electronic CAD and the engineering workstation. The lecture content looks at the way CAD facilities have evolved from centralised batch-processing main-frames into networked workstations, each with a high level of computing power but with distributed access to file systems. The core material in subsequent lectures takes an example drawn from an earlier course in digital systems given the previous term, and analyses it from a somewhat different point of view, by decomposing it hierarchically and considering which simulators can be used at each level of abstraction.

In the earlier course, students are introduced to such building blocks as n-bit adders, n-bit registers and comparators, each built up from single-bit cells[1]. These modules are then used in a bottom-up implementation of a module which, on command, produces a pulse of given length, its duration determined by a set of binary switches. This bottom-up approach is conceptually easier to comprehend than a top-down hierarchical design exercise based on a specification of the required circuit. However, it is precisely the latter approach which the ECAD lecture course examines—it takes the specification of the module and beginning with the top-level architecture (see Fig. 1) decomposes the design. At each level of abstraction, the suitability of behavioural, functional, gate and circuit level simulators is discussed and examples given.

At this point it should be stated that the emphasis is not on the methodology of top-down design, since first year undergraduates are simply not knowledge-able enough in the building blocks available at the lower levels of abstraction. The emphasis is rather on the way a full design can be structured with each level hiding the complexity of levels lower down. The similar approach in structuring programming is referred to.

In the context of the pulse generator design example, hierarchical schematic capture is explored using Silvar Lisco's structured design system (SDS) and in

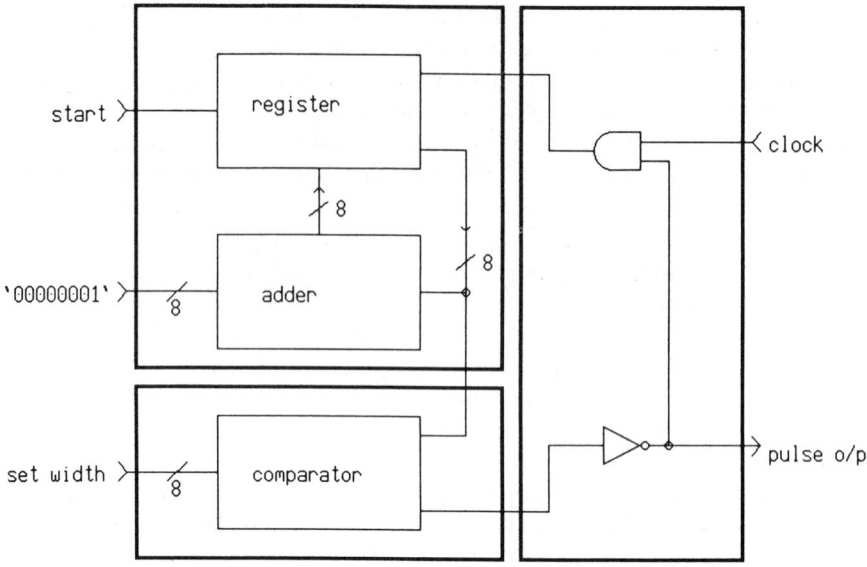

FIG. 1   Pulse generator.

particular CASS (Computer Aided Schematic System) as a concrete example. The major characteristics one expects from most circuit capture systems are highlighted, such as the availability of libraries, the generation of user defined symbols, the menu-driven interface, the plotting facilities and the availability of interfaces to other standard design systems.

The behavioural simulator HELIX, the logic simulator BIMOS, both from Silvar Lisco, and the circuit simulator SPICE are all looked at in the context of simulation of elementary circuits. The two phase clock generator shown in Fig. 2, is one such example. Based on just NOR gates and an INVERTER, the circuit is simple enough for complete analysis in the course of a lecture using both HELIX and BIMOS, but interesting in its own right. The description of both an inverter and a nor gate in HELIX's Pascal like Hardware Description

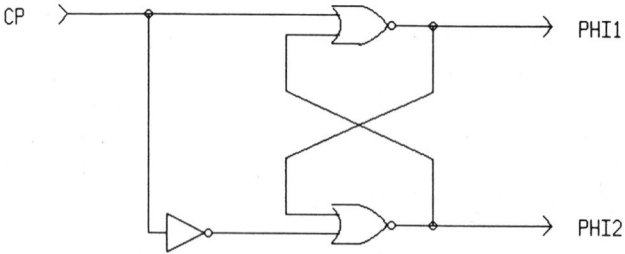

FIG. 2   2-phase clock generator.

Language (HDL) is not too complicated a demonstration and serves as an adequate example of using an HDL. Using BIMOS is simpler and a typical command file is illustrated in the course of a lecture.

The final lecture is devoted to the circuit simulator, SPICE. In particular, its use is illustrated in the analysis of passive filter networks, and other simple circuits of a complexity which undergraduate students could, in principle, analyse with pen and paper. In this way the use of CAD as a **useful** tool to the student is emphasised.

In summary, the course provides a CAD context which can be employed by other lecture courses. For example, circuit-orientated courses can refer back to the use of SPICE for circuit analysis.

Finally in this section, it is also of interest to note that finding suitable texts for the course was an unenviable task. No single text covers all the course material adequately. However three books are recommended for background reading: *Computer Aided Electronic Engineering*[2] is a good overview and covers material such as computer aided manufacture and testing, *CAD for VLSI*[3] is advanced but has an in-depth coverage of simulation at various levels of abstraction, and, *An Introductory Guide to Silvar Lisco and Hilo Simulators*[4] is just that—a guide to using the simulators examined in the course.

4   WORKSHOP CONTENT

The main aim of the workshop is to introduce the student to some of the features of an advanced electronic CAD environment, in this case the Silvar Lisco SL2000 system. A 4-bit asynchronous counter is used to illustrate the process of entering a design in schematic form, and then running a simulation to check the circuit's logical behaviour. Separate schematics are entered for the ripple counter and the (simple) decode logic and the top-level design entered as a third schematic sheet. In Fig. 3 the unit RIPPLE represents four J–K flip flops whilst within the unit DECODE there is a 4 input AND gate, the output of which is used to reset the counter. The students can subsequently modify the interconnections in DECODE to generate different value counters. Thus not only do the students get accustomed to using the schematic editor, but also adapt a methodology which demands that symbols be created representing

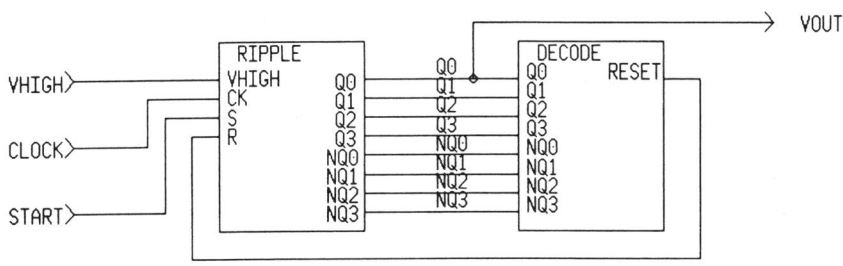

FIG. 3   Asynchronous counter.

parts of the circuit, which can subsequently be used at a higher hierarchical level.

The workshop script is a stage-by-stage instruction manual for performing the required operations on the workstation. Once the schematics have been entered, and the design saved, the necessary steps to simulate the design and view the results are listed. Before a simulation can be performed the net list extractor, NLE, and the hierarchical design expander, HIDEX, both part of the structured design system, must be run, and a command file for the simulator BIMOS prepared. Students are asked to use the screen editor to copy the relevant commands from the workshop script. Errors in the circuit do not usually come to light until the simulation is run. At this stage it is a laborious task correcting any errors due to incorrect entry with the schematic editor and the workshop staff usually need to be at hand to guide the students through a successful simulation.

At key stages of the tutorial, students are asked to plot the results. Plot files are easily generated by pressing a single keyboard key and a screen dump is obtained on the laser printer.

A second exercise forms the remainder of the tutorial. The aim here is to enter and simulate a modulo-16 synchronous counter without any specific help from the workshop script.

Assessment is on a fairly simple basis. It is based on attendance records, the time-stamped plots obtained during the tutorial and performance assessment by workshop staff.

Because part of the aim of this course is preparation for subsequent use in project work, the tutorial script also includes information on the operating system. The way window adjustments are performed, and files edited and read is explained. Simple commands such as those which help the user to manipulate directories and files are listed. The script thus serves as a self-contained introduction to any persons wishing to use the system, provided they have logins.

5 OBSERVATIONS

We have noticed that students with prior experience of a workstation environment usually complete the scheduled experiment with considerably greater ease than those with no such experience. Hence, by providing the students with an opportunity to work in the laboratory, they are ready to undertake similar work on different systems without having to overcome large learning curves. Because they have been introduced to the available facilities so early in their studies, subsequent use of the tools needs little re-learning. We have yet to see what the impact of this introductory course will be on final year projects, but envisage a much greater demand on the available facilities next academic year when, for the first time, students who are already familiar with the tools will use them for project development and documentation.

The use of SPICE for circuit analysis in the context of circuit orientated courses has been referred to above. The advantage of a formal lecture on the topic is born out by the improved performance in the second year practical on SPICE, which is now completed in about 75% of the time taken in previous years.

User feedback has always been of particular importance in the area of computer aided design, and in itself highlights the great advantages of the availability of source code and full software support staff. Unfortunately in the case of commercial software, source code is not available and user change requests such as an improved man-machine interface or better feedback between editor and net-list extractor, can rarely be acted upon.

One of the most common questions asked by our students is when do they use the ECAD laboratory again. Until recently, the facilities have been used largely by final year students for their projects. In an attempt to introduce a more formal structure to the use of the laboratory throughout a student's three year course, a second year workshop in VLSI design is in the process of being prepared which will come into operation in the 1988/89 academic year. The workshop will be based on the use of an integrated, menu-driven set of design tools, obtained from industry and written in-house, for the design of integrated circuits. The same tools have been used for some time by final year students in chip design projects[5].

6 CONCLUSION

The use of electronic CAD tools is nowadays as much a fundamental engineering requirement as the use of oscilloscopes, voltmeters, logic analysers and soldering irons. Thus it needs to be taught at the same time as the use of other fundamental tools. The introductory ECAD course and workshop attempt to address this requirement.

7 ACKNOWLEDGEMENTS

We acknowledge the help of the University Grants Committee and GEC Avionics, Rochester, in helping to finance the establishment of an ECAD laboratory at the University of Kent.

8 REFERENCES

[1]   Dimond, K. R., *private communication*, University of Kent (1987)
[2]   O'Reilly, W. P., *Computer Aided Electronic Engineering*, Van Norstrand Reinhold (1986)
[3]   Russell, G., Kinniment, D. J., Chester, E. G., McLauchlan, M. R., *CAD for VLSI*, Van Nostrand Reinhold (1985)
[4]   Blundell, B. G., Daskalakis, C. N., Heyes, N A. & Hopkins, T. P., *An Introductory Guide to Silvar Lisco and HILO Simulators*, Macmillan (1987)
[5]   Walczowski, L. T., Waller, W. A. J., Dawson, C. & Dimond, K. R., Student Full Custom Design, *IJEEE*, this issue, p. 113 (1989)

ABSTRACTS–ENGLISH, FRENCH, GERMAN, SPANISH

Introducing electronic CAD to first year students
An introductory course in Electronic CAD for first year students has been developed. The course gives basic training in the use of CAD tools and looks at their role in the design of an electronic system. This paper describes the course, its origins and our experiences in giving it.

Introduction à la CAO électronique aux élèves de prèmiere année

Un cours d'initiation à la CAO électronique a été créé pour les étudiants de première année. Ce cours donne un premier contact aux outils CAO et montre leurs roles lors de la conception d'un systeme electronique. Ce papier décrit le cours, son origine et notre expérience en le donnant.

Einführung von erstjährigen Studenten in computergestützten Elektronikentwurf

Ein einführender Kursus in computergestützten Elektronikentwurf für erstjährige Studenten wurde entwickelt. Der Kursus bietet Grundlagenausbildung in der Benutzung von CAD-Rüstzeug und betrachtet dessen Rolle im Entwurf eines elektronischen Systems. Die Arbeit beschreibt den Kursus, seinen Ursprung und unsere Erfahrungen bei seiner Darbietung.

Introduciendo CAD electrónico a estudiantes de primer año

Se ha desarollado para estudiantes de primer año un curso de introducción al CAD electrónico. El curso da entrenamiento basico en el uso de herramientas de CAD, y examina el papel que desempeñen en el diseño de un sistema electrónico. Este artículo describe el curso, sus orgenes y muestra experiencia con el.

INTRODUCING STUDENTS TO ELECTRONICS COMPUTER AIDED DESIGN

J. T. PROUDFOOT and P. A. MAWBY
Department of Electrical and Electronic Engineering, University of Wales, Swansea

1 INTRODUCTION

Swansea University runs a broad-based degree scheme in Electronic and Electrical Engineering. Students on this scheme follow a fixed syllabus for their first two years, when the fundamentals of the subject are presented. In their third year, students may specialise, by choosing five options from the range on offer (typically ten) together with an individual project.

Prior to the Electronics Computer Aided Design (ECAD) initiative in the UK, the use of computers as engineering tools was met only through the teaching of digital systems programming and microprocessor work, and by a few students using proprietary systems for their project work. The 'inadequate funding' noted by Powner[1] applied here took, and it was only through the ECAD initiative that sufficient facilities were made available at a reasonable cost to allow all students to be introduced to what is generally understood by CAD, to assist in the design of circuits and systems. The need nowadays to teach students about the design of custom chips or application specific ICs (ASICs) and the case for supporting this, as other engineering teaching, with practical work, leading to a final product (in this case, working silicon) has been well made elsewhere[1,2,3,6]. The interest shown by prospective employers whilst interviewing students who have taken our ASIC design course reinforces the pertinence of this subject nowadays; there is no doubt that these students realised they had acquired a very useful 'string to their bow'. The ECAD initiative, therefore, was timely in providing the CAD tools essential in ASIC design, and also in setting up cheap routes for the fabrication of silicon.

The initiative enabled two activities to be included in our degree course from the 1987–88 session. One was a p.c.b. design exercise as part of the students' EA1* training, whilst the other was the ASIC design course which is described in the remainder of this paper.

---

*Engineering Applications, as recommended by the Committee of Inquiry into the Engineering Profession, London, chaired by Finniston, H.M.S.O. Cmnd. 7794 (Jan., 1980).

2 COURSE STRUCTURE, CONTENT AND ASSESSMENT

For reasons similar to those given by Hardy[2] our ASIC design course was introduced as a third-year option. We felt that, earlier than this, students would not have the requisite knowledge to apply to their design. This way, too, we were able to limit the numbers taking the course to 20, to suit the practical facilities available (a network of 10 Apollo DN330 workstations).

Since this was to be an optional course, it was important that it presented the same workload to students as other options; up until now this had been exclusively 48 hours of lectures. However, as has been noted previously, it is imperative to support this material with practical work, so some precedent had to be devised to mix lectures and practical work equivalent to 48 hours of lectures. The formula chosen here was to present 26 hours of lectures and provide 33 hours of supervised practical sessions. Using a weighting of '$\frac{2}{3}$ practical hour' = '1 lecture hour' this equated to the 48 hours required. Students were expected to put in further hours of practical work themselves, equivalent to tutorial work and background reading time in other subjects and indeed most did spent considerable additional time in the design room.

2.1 *Lecture content*

The principle underlying the course was to introduce students to design techniques and methodologies relevant to the efficient production of working silicon, using available commercial tools and facilities. The emphasis was on the use of a realistic tool to do a job of engineering. There was no attempt formally to compare CAD suites, although some commonly available features, not available in the suite used, were described. Similarly, there was no description of the way in which CAD software should be designed. Although we acknowledge that these topics are relevant to an ASIC designer wishing to make an assessment of his 'best' CAD suite, the time available to the present course precluded them. To ease the introduction of the new course, it was made up from a number of units, some of which were transferred from existing courses, and given by a number of staff members.

The course was based around MCE's (Microcircuit Engineering Ltd.) BX gate-array design suite. This is self-contained and relatively straightforward to use and students had little difficulty learning to use it. Gate-array implementations are also the shortest route onto silicon and are relatively easy to understand for first-time users. In a course of this length these are important considerations[1,2,3,7,8]. Some lectures were, of necessity, devoted to describing how to use BX. Included in this unit, however, were lectures on design philosophy, modularity and the principles of simulation at different stages of the design. Also included were lectures on the characteristics of the microelectronic processes used for the implementation of the ASIC. Some knowledge of these was seen as essential to understand the design rules and constraints imposed by the CAD software. Indeed Morant[4] and Waller[8] state that this knowledge is important and that full-custom design is the most advantageous educationally even if, subsequently, the CAD software hides much of the complexity. However, they also note the complexity of full-custom software, and the need for students to learn it easily, noted earlier. Given the length of our course and its level, we felt justified in using gate-array technology. This unit was covered in 8 hours of lectures although since it was given by the authors, who also supervised the practical work, the material was expanded in tutorials during the practical sessions.

Another unit, of 9 hours, was devoted to 'Design for Testability' covering such issues as partitioning, scan-path analysis and test vector generation. A further unit, also of 9 hours, covered 'System Design Engineering' as an introduction to the essential topic of formal system specification. Whilst, at present, this is more theoretical than we would like, we hope in time to integrate it with the CAD tools in use.

2.2   *Practical work*

This was split into two sessions. In the first term the students had 10 hours, working in pairs, to explore a simple design to enable them to become familiar with the BX software. This work was not assessed. The design set was a simple sequence generator; the expected solution was based around a counter. In the event, several solutions were proposed and as the purpose of this session was exploratory, these were pursued and results compared at the end of the session. Whilst this was interesting educationally, we will see later that the flexibility probably detracted from the success of subsequent work.

In the second term the students were grouped in fours and had 20 hours (supervised) to work on their main assignment. Waller[8] notes the advantages of group design work viz. larger, more realistic projects are possible with, consequently, better student exposure to more aspects of a design and the experience of working as a team member which, nowadays, is considered an important part of engineering education.

The assignment set was to produce a d.c. motor-controller, using pulse-width modulated drive and opto-coupler-generated speed feedback. An experiment board for a microprocessor laboratory was available, giving a motor, opto-coupler speed-demand switches and diagnostic LEDs, all with TTL interfaces, thus simplifying the student's problem to the production of the controller chip. An added bonus of this was that we were able to provide and demonstrate to the students a microprocessor-based solution to the assignment. This gave them some insight into the count and comparison operations that were required in the controller. A basic solution requires four counters of 6 or 8-bit width plus two comparators and some control and diagnostic logic; in all about 800 gates. The counters vary in their requirements; one is up/down, some require presetting, another needs only to be cleared. This set the students the dilemma of whether to create one large, general-purpose counter module and repeat it throughout the design, thus saving on design time, or whether to tailor each counter to its specific needs to reduce logic. Both approaches were adopted by different groups with comparable levels of success being attained; certainly the former method is easier to assess and is correspondingly probably preferable unless space is at a premium.

It was our intention to fabricate one selected design over the Easter vacation and to spend the final practical hours at the start of the summer term commissioning the ASIC. Each group was expected to build a simple board for the controller and its interface to the experiment board. In the event, as we will discuss in the following section, no designs were considered suitable for fabrication by Easter, so this part of the course was not completed in this year.

One of the students on this year's course chose as his project the design of a tester for MCE gate-arrays. The tester was PC-based and took simulation and placement files in standard BX format, and translated and applied them to the gate-array via a simple interface card for the PC, reporting errors where appropriate. The tester could not, of course, work in real time, but logic levels and sequences were maintained as per the original design simulation. This allows the delivered ASIC to be verified against the supplied design with a significant level of confidence. The tester works well, and will be available in future years as part of the commissioning process.

2.3 Student progress

As we have stated already, no group completed their design to a standard ready for fabrication, although three groups produced designs which, on examination, had only minor flaws. Mostly these had all proven modules of their design, but had been unable to verify the overall top level of the design. Any of these could form the basis of a submitted design with only minor alterations. Two main causes for this failure to complete can be examined and our experience is reinforced by that of others in the early years of similar courses[2,3,4,5,7,8]. Firstly, there is some feeling that our assignment was too complex, and certainly it appears so in comparison with some assignments on courses of a similar level. However, it was a group project and so could be expected to be more demanding. The students did find great difficulty, though, in formulating the final overall simulation of their closed-loop control system. We believe, however, that attention to the second cause will be more productive. As noted earlier, our first term practical work, whilst valuable educationally, was of little material benefit, in most cases, to the main assignment. The earlier references have noted the need for strict monitoring, timetables and benchmarks in the development of student silicon and that students are notoriously poor in their attention to the detail that is required to complete a design successfully and our experience has concurred with that. Thus we believe that by ensuring that our first-term familiarisation exercise is directly applicable to the assignment (e.g. a general-purpose counter module) and by closer monitoring of the detail and progress of the assignment, made possible by our increased experience, it should be possible to bring more designs to fruition.

Despite the problems, however, there is no doubt that our students enjoyed the course and, as noted previously, recognised its value to them. Their motivation reflected that of others[2,4,8], as demonstrated by the long hours spent in the design office (perhaps also indicative of an overdemanding assignment)!

2.4 *Assessment*

The precedent set in establishing an option containing practical work extended to the assessment of that option. We felt that the practical work must be assessed, reflecting its importance to the subject and to give it credibility. Furthermore, whilst some of the practical experience could be examined by written questions, the reduced lecture content meant it would be difficult to set a normal full 3-hour written exam and maintain standards. In the event, we decided on a 2-hour written examination worth 60%, with the other 40% for the option coming from our assessment of the second-term assignment, this being a reasonable reflection of the course structure. We were very much aware that these students would be gaining the equivalent of two written questions of their colleagues on other options, from a subjective assessment of their contribution to a piece of group work. Carter[6] notes the problems of quantifying student qualities in project work and the view that it is questionable if such marks should contribute to degree classifications. We were doing precisely that, but justified it on the grounds that, although the qualities we were assessing were very different from those required for a written examination, they were, nonetheless, important in a subject deemed worthy as an option in the final year of a degree course. However, we remained aware that, for the first time in our experience, such an assessment was not being applied to the whole class, but could influence, fairly or unfairly, the subset who had chosen this option, and that our assessment, as well as being as equitable as possible, should return marks comparable to those of written questions. We were aware, also, that good individuals could be penalised by being grouped with weaker colleagues (and vice versa) and tried to watch out for such occurrences.

To assist with our assessment of the practical work we drew on our experience in second-year digital systems laboratory group-design projects, which had been running for some years. A detailed notebook was maintained throughout the supervised practical sessions, noting group progress, where assistance was required and the contributions of individuals to the group effort. An oral examination was carried out of each group where, after submission of documentation describing their design, they could be questioned on points of detail they had contributed to the design. These proved to be lively and very enjoyable (if exhausting) sessions. Much can be learned of a student's character in such situations from their willingness or otherwise to defend or criticise a colleague! A checklist of salient features of each student's approach to the project, similar to that given by Carter[6] as an assessment schedule for experimental investigations, was used to quantify the individual's marks. In the event, there was little correlation overall between the marks scored by our students on their examination paper and from their practical assessment. Given that these assess different qualities, as we have noted, this is perhaps not surprising. However, with a small group of just 20 students, it was possible for us to review each case individually and ensure that each had neither benefitted nor been penalised unfairly. Hardy[2] and Carter[6] note that such supervision and assessment is very labour-intensive but is, nonetheless, an enjoyable and correspondingly rewarding form of tuition, and we would concur with those sentiments. Carter goes further to make some very pertinent, if provocative, comments on the unattractiveness of such teaching duties to the ambitious academic who places more value on his research time. In the light of the current requirement for group design work in undergraduate engineering courses, such comments demand some attention.

3   CONCLUSIONS AND COURSE DEVELOPMENT

The ECAD initiative has made it possible for us to mount a new course in the important area of ASIC design. The course has been reasonably successful in its first year of operation but will be 'tuned' as has been discussed, for future years. In the longer term we hope to be able to expand the course to cover the existing topics in more depth and to introduce new ones, particularly technologies other than just gate-arrays. We fully expect an increase in the amount of project work, making use not just of the CAD tools, but of the now cheaply-available silicon, and investigating the characteristics of these devices themselves. This must increase as our colleagues recognise the viability of the processes available.

ACKNOWLEDGEMENTS

The authors would like to acknowledge the contributions of their colleagues Mr L. A. M. Bennett, Dr R. Artym, Dr J. S. Mason and Professor H. A. Barker, made during discussions about the format of this course.

REFERENCES

[1]   Powner, E. T., 'Microelectronics education' (Editorial) *Int. J. Elec. Eng. Educ.* **22**, No. 1, pp. 3–4 (1985)

[2]   Hardy, C. J. et al., 'Teaching in silicon', *ibid.*, pp. 5–11

[3]   Hurst, S. L., 'Low-risk ... semi-custom microelectronics for undergraduate teaching purposes', *ibid.*, pp. 13–20

[4]   Morant, M. J., 'Integrated circuit design in an industrially-oriented M.Eng. course', *ibid.*, pp. 29–36

[5]   Cottrell, R. A. et al., 'Fast-turnaround gate array committment using E-beam generated masks', *ibid.*, pp. 37–49

[6]   Carter, G. et al., 'Assessment of undergraduate electrical engineering laboratory studies', *IEE Proc.* **127** PtA No. 7, pp. 460–474 (Sept., 1980)

[7]   Harrison, G., 'Redcad user experience', *ECAD Lead-Site Liaison Meeting, Huddersfield* (April 1988)

[8]   Waller, W., 'Undergraduate full custom design', *ibid.*

ABSTRACTS–ENGLISH, FRENCH, GERMAN, SPANISH

Introducing students to ECAD

This paper describes how ECAD was introduced into a typical Electrical Engineering degree course. The structure of the ECAD module is presented, together with its integration into the course. The module exercises are described and the paper concludes with observations of their success, together with planned developments for future years.

Introduction à ECAD
Cet article décrit comment ECAD a été introduit dans un cours de la spécialité Ingéniérie électrique. La structure du module ECAD est présentée, avec son intégration dans les enseignements. Les exercices du module sont décrits et l'article en conclusion observe leur succès et prévoit les développements des années futures.

Einführung von Studenten in ECAD
Dieser Artikel beschreibt, wie ECAD in einen typischen Graduierungskurs für Elektrotechniker eingeführt wurde. Die Struktur der ECAD-Module wird vorgestellt einschließlich der Integration in den Kurs. Die Modulübungen werden beschrieben und Beobachtungen über deren Erfolg zusammen mit geplanten Weiterentwicklungen für die zukünftigen Jahre einbezogen.

Introducción de los estudiantes en ECAD
Este articulo describe como se introdujo ECAD en un curso de grado de ingenieria electrónica. Se presenta la estructura del módulo ECAD junto con su integración en el curso de describen los ejercicios de módulo y concluye con observaciones sobre su éxito junto con futuras aplicaciones para próximos años.

THE DESIGN OF CMOS GATE ARRAYS AS STUDENT UNDERGRADUATE PROJECTS

T. M. McGINNITY, S. G. JONES, M. DOWDLE and K. BIRKINSHAW
Department of Physics, The University College of Wales, Aberystwyth, Wales

1 INTRODUCTION

The response of the University Grants Committee in the United Kingdom to the proposals of its working party on Electronic Computer Aided Design in education (ECAD)[1] in 1985 was to make a limited amount of funding available to universities for the purchase of hardware platforms on which to run commercial grade software. In addition, a range of such software was subsequently purchased centrally on behalf of the universities and polytechnics and has recently been substantially extended. Currently the scheme includes, among others, packages such as the Silvar-Lisco SL2000 suite, HILO 3 from Genrad, MCE's BX Design package, ELLA from Praxis and ISIS from Racal–Redac.

At the University College of Wales, Aberystwyth, the possibility of incorporating the computer aided design (and subsequent fabrication) of integrated circuits into the degree course in Microelectronics and Computing (MEC) was greeted with enthusiasm. This degree programme is an integrated course run jointly by the Departments of Physics and Computer Science. Although a course in Integrated Circuit Design was already part of the degree scheme prior to this, no facilities existed for any associated practical work, and the relevant staff were of the opinion that such a situation was educationally unsatisfactory. There was general agreement that any course in integrated circuit design should have, as an integral part, not only associated compulsory sessions in a computer aided design laboratory, but also the opportunity to complete the learning cycle by having at least some student designs fabricated and returned for testing. The opportunity to acquire high quality hardware and software could not be ignored.

A decision was taken to purchase Apollo workstations (DN560 and 330), despite the fact that a considerable amount of funds had to be added to those made available by the UGC. The workstations were in place by January, 1986. Throughout the summer of 1986, staff from this college attended courses run by Silvar-Lisco on such packages as CASS, HELIX, GARDS, CALMP and PRINCESS, in order to familiarise themselves with this software. Experience was also gained in HILO during this period.

This paper attempts to assess the results of implementing ECAD work within the Microelectronics and Computing degree course at Aberystwyth.

2 INCORPORATION

The actual introduction of ECAD practical work into the degree scheme occurred for the first time in the academic year 1986/87. This was achieved by introducing a new 8-lecture course on two main topics into the student's second year. These were (a) the use of Silvar-Lisco software as tools for hierarchical design and (b) a methodology for the design of testable gate arrays. On the grounds of both cost and available student time, a decision was made to concentrate practical work solely on gate array design. However, a 10-lecture course on MOS full-custom IC design was also established and runs in the final year. (The extension of this course by offering a further 10 lecture course as an option is under consideration.) Silvar-Lisco software tools were chosen purely on grounds of availability and lecturer familiarity at that point in time. The wisdom or otherwise of such a choice is discussed later in this paper.

As part of their course, each student was required to design, as a project, a circuit to be implemented on a Texas Instruments TAHC gate array. The objectives of such work were to provide the student with experience of using at least one commercial ECAD software package to perform gate array design, to introduce him/her to some of the problems which a designer encounters while developing an integrated circuit and finally to impress upon the student the limitations of currently available software and hardware. Students were encouraged to adopt a critical yet constructive approach. The boundary conditions were that each design should be of the order of 250 gates (2-input NAND equivalent) in size and utilise 12 or fewer I/O pins. The number of students on the course allowed for a commitment by the Departments to meet the costs of fabricating those student designs which met certain minimum standards, as assessed by the supervisor. The TAHC10 array was selected, with the design process based firmly on the SDS–CASS–HELIX–GARDS route. Three students would share an array. Fabrication of designs was to be performed by a company called Qudos, which offered fast turnaround times from receipt of design, subject to certain volume requirements. Students were free to propose their own project topic, but failure to submit a project title plus 1-page summary by the deadline led to an imposed subject. It is important to note that it was the actual design experience rather than the design complexity which was of interest, although novelty of topic was encouraged. However the limitation on the number of gates mitigated against anything very sophisticated. It is intended to discuss two of the student projects which reached fabrication, not on the grounds of truly innovative design but because they allow for an assessment of student response to ECAD work as part of their degree scheme.

3 DISCUSSION OF PROJECTS

The first of the designs under consideration was entitled *A shaft encoder with integral seven segment display driver*. This project aimed to design an integrated circuit which would accept two signals from transducers and determine from these both the speed and direction of rotation of a shaft. Denoting the two input signals as A and B the requirement was that A and B were to be 90 degrees out of phase with respect to each other. Thus, assuming the shaft were to rotate in a clockwise direction, a pulse arriving at A before the corresponding pulse on B, would indicate clockwise rotation and the counter would increment. Should B arrive before A the counter would decrement. The circuit was also to be able to handle a change in direction. Thus sequences such as ABABAB indicate clockwise rotation, BABABA indicate anticlockwise rotation, and ABABBBABA indicate two clockwise revolutions followed by two anticlockwise revolutions. Clearly direction changes in this scheme could generate small errors. However for the purpose of this project these were not considered important. The circuit was also to be able to directly drive a three digit seven segment display.

The design was approached in a hierarchical manner, being broken down into blocks each of which was subsequently broken down into smaller blocks until a gate level was reached. The overall design is shown in Fig. 1, where it is seen to consist of several functional blocks, some of which are identical. The central part of the circuit is the chain of three up-down counters, one for each of the digits required. The counters are identical and are cascaded. Since the circuit output was required to be in decimal, the counters count using binary coded decimal, to avoid the requirement for a binary to decimal converter.

The direction decoder, labelled 'Pulse' in Fig. 1, takes a pair of negative-going inputs which are in quadrature and produces one of two outputs depending on the direction information inherent in the inputs. While the operation of this sub-circuit appeared initially to be satisfactory, as evident from the results of a HELIX simulation, subsequently problems were encountered where the counters appeared to get stuck in an undefined state. This was traced to the input pulses being of insufficient width, leading to intermittent errors, particu-

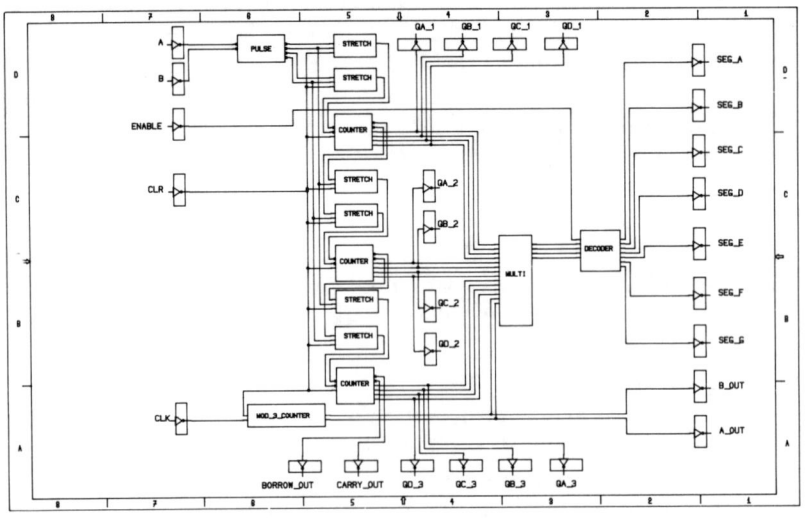

FIG. 1   Block diagram of the shaft encoder project circuitry.

larly when counting down. To ensure reliable clocking of all stages of the counters, the student found it necessary to 'stretch' the input pulses. This was achieved via the blocks labelled 'stretch' in Fig. 1.

The multiplexor block takes the outputs of each of the counters and multiplexes them onto the bcd to seven segment display decoder. It is composed of four one of three multiplexers. The bcd to seven segment decoder is a standard design, and incorporates display blanking via the ENABLE input. Hence it is not possible to blank a single digit, but this was not thought to be important in this application. The mod three counter is required to select which of the inputs is to be connected to the decoder and its output thus drives the select lines on the multiplexor.

The individual blocks of the circuit were simulated using the HELIX simulator. Results achieved indicated that the design should function correctly. The complete design, excluding pads, was then simulated and also indicated correct operation. Tentative experiments to find the maximum speed of operation indicated a value of 3MHz, which was accepted as satisfactory.

In an attempt to make the design more testable and localise faults, it was decided, as far as possible, to utilise unused pins to bring out connections to each major sub-block of the design. This was initially facilitated by a decision of the project supervisor to make a full TAHC10 array available for this project, as opposed to the original one third. It was recognised that this procedure would not normally be available in a gate array design.

The design proceeded to the layout stage, using the Silvar-Lisco GARDS product, specifically GPLACE. This is an automatic placement program, with the possibility of manual intervention. Initially, the program was given a free rein to place all components with the exception of some pads, using its 'constructive' placement mode. Subsequent attempts to route the design failed to achieve 100% routing, as expected. Accordingly a more time-intensive, but reliable, approach to placement was adopted, whereby the overall design was partitioned into placement blocks, each block being allocated specific space on the array. Within each block, placement was made automatic. The placement program was then used to improve the layout based on the criteria of gate interconnect length and congestion. The optimisation program was repeated until less than 1% improvement in interconnect length and less than 5% improvement in congestion was being achieved. Figure 2 illustrates the actual utilisation of the array. The upper right corner area was reserved for use by Qudos for testing purposes.

Routing of the design was achieved using the Silvar-Lisco automatic router SLROUT. The only guidance the program was given was to attempt to route in four waves, starting with the shortest interconnect lengths. The routing program was unable to make 28 of the 495 connections. In an attempt to solve this problem a number of the connections brought out solely for testing were sacrificed. Despite this, three connections remained incomplete at the end of the routing phase. It was found necessary to make these connections manually. This operation took a considerable period of time. A colour monitor was

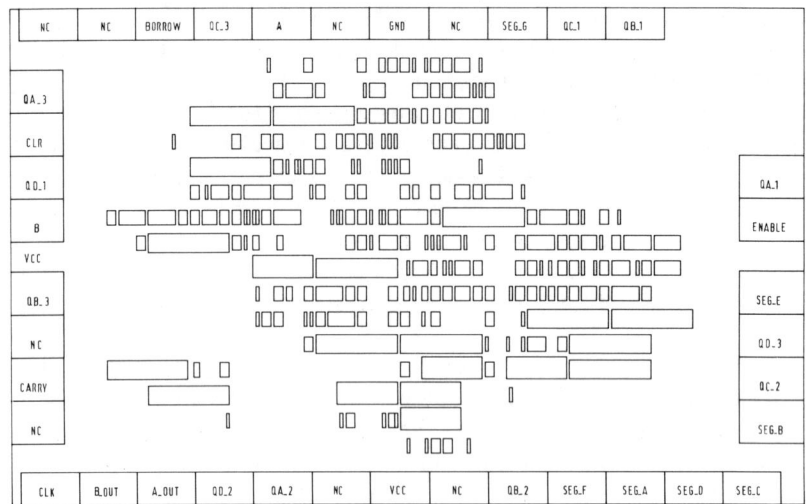

*FIG. 2   Placement map of the shaft encoder project on a TAHC10 array,
indicating cell utilisation.*

mandatory for this part of the work. Even with such a facility the students
found manual routing of interconnects extremely difficult.

The next logical step was to perform back annotation and re-simulate the
placed and routed design. Unfortunately this proved impossible with the
software available and the design had to be sent for fabrication without this
crucial step being performed.

The second project was a more modest affair and set out to implement a
cascadable 6-bit binary to bcd converter. This project, while initially intended
to occupy one third of a TAHC10 array, was actually implemented on a
TAHC06 array. The design uses the familiar 'addition of three' algorithm,
whereby the binary number, most significant bit first, is clocked into a shift
register. If, after a clock pulse, the binary number in the four least significant
places of the shift register is greater than or equal to binary five, then binary
three is added. Otherwise nothing is added. After six clock pulses the conversion is complete and the values held in the shift register represent the bcd
number. The task was, therefore, simply to produce a shift register, an adder
and an element of circuitry designed to test whether the number currently held
in the four least significant bits is greater than or equal to binary five. A block
diagram of the device is illustrated in Fig. 3.

The procedure for the design was essentially the same as for the project
described above. However, the reduced complexity made the problem significantly easier. After using SDS and CASS to capture the schematic, the design
was simulated using HELIX, commencing with the sub-blocks and moving on
to the complete design. Time limitations prevented simulation for all possible
inputs but the successful results obtained for those codes used were taken as
indicative that the circuit would probably work correctly overall.

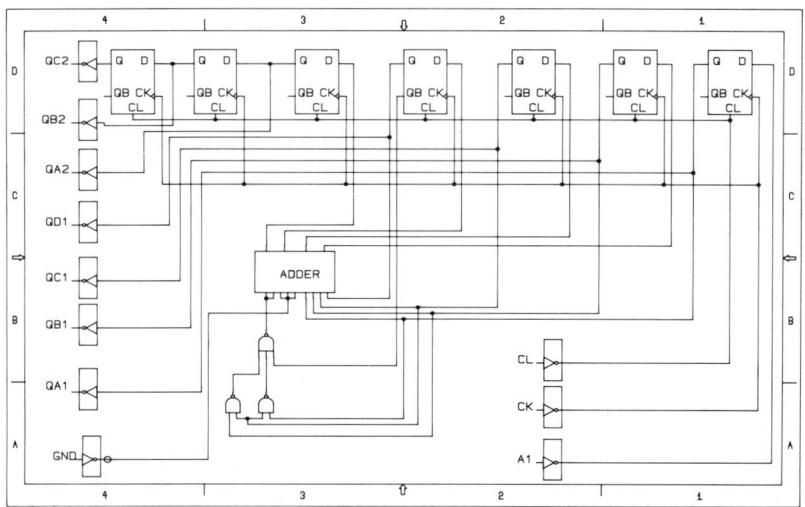

FIG. 3   Block diagram of the binary to bcd converter project circuitry.

The design was placed and the placement optimised in a manner similar to that described previously. Despite the target array being the smaller TAHC06, 100% routing was achieved without manual intervention. This was clearly due to the smaller number of interconnections required (109 as opposed to 495). As before, no back annotation was possible. The design was converted to GDSII format using the Silvar-Lisco programs ARTWORK and MASKOUT and sent for processing.

4   TESTING AND RESULTS

Both arrays discussed above were fabricated. Five devices per design were returned for testing. The test strategy in both cases was rather simple and, essentially, consisted of designing and building a test circuit to check functionality of the device.

The shaft encoder was tested first and was found only partly functional. It was unresponsive to input signals, regardless of frequency. It was possible, however, to confirm that the mod-3 counter was functioning correctly, by examining the output pins attached to A_OUT and B_OUT. The display could be cleared and set to 000 by asserting the CLEAR line. The input circuitry which decoded the phase information into a direction signal appeared to be non-functional. Unfortunately, the outputs of this section were not connected to external pins and so it was not possible to examine the outputs or, alternatively, provide the correct signals in order to test the counters. All five devices produced the same result. It is probable that the inability to extract the various capacitances from the placed circuit and re-simulate, was the root of the device failure.

The binary to bcd converter was tested in a similar way. A test circuit was built and all possible input combinations were examined. In this case correct functional behaviour was observed. This was fortunate, as no real provision had been made for assessing the operation of internal nodes. Limitations on time were the major factor in this omission, but, clearly, some provision for fault localisation should have been made. There is no doubt that one of the principal advantages of having designs fabricated is that it leads to a much greater awareness of the need to seriously consider a testing strategy at an early stage.

5 DISCUSSION

At this stage of experience, having instituted courses and project work which have resulted in the actual fabrication of gate arrays in an undergraduate educational context, it is beneficial to review the problems apparently inherent in such work, assess student response and examine again the merits of maintaining ECAD projects within a modern electronic engineering curriculum.

There are clearly a number of very positive aspects to the situation. Of major importance is the acceptance by both academic staff, students and employers that familiarisation with computer aided design techniques, modern workstations and commercially available software is of crucial importance for graduating engineers. A number of students at Aberystwyth have found that ECAD experience has arisen as a topic of discussion at interviews. Furthermore the MEC course at Aberystwyth involves a year's experience in industry, between the second and third year of study. A considerable number of students have found themselves engaged in ECAD activity during such an industrial year. There appears to be no doubt that, as the market for ASICs grows, the case for inclusion of ECAD materials in degree schemes will be irrefutable.

Student reaction to ECAD work is also of interest. Predominantly, MEC students at Aberystwyth have expressed satisfaction at the incorporation of such work into their degree scheme and have thoroughly enjoyed designing an integrated circuit, despite occasional mutterings about extended simulation times. The possibility of having their designs fabricated has been an important motivating factor. One problem which has concerned them, however, has been the time required to take a small design from concept to fabrication.

It is true to say that the academic staff involved with ECAD work at Aberystwyth are more convinced than ever of the need for such project work or similar activity. From an educational point of view, ECAD work offers the opportunity to students not only to develop their skills in CAD but also to improve their basic understanding of digital design, circuit operation and faultfinding in general. One area which needs more emphasis is that of testability.

There is, of course, the negative side. From a student point of view, with software such as SL2000, there is a very long lead-in time during which one is attempting to gain familiarisation with a large number of different packages. In retrospect, the Silvar-Lisco software may appear somewhat daunting for a student, simply by virtue of the fact that it is tremendously flexible and offers a wide number of options at almost every step. This, presumably, arises as a result of not being tied to a particular fabrication route. A simulator such as HELIX is probably over-powerful for the level of student designs which one might expect to encounter in an undergraduate programme, other than a major final year project. Given sufficient time, this is not, of course, a problem. However, in most engineering courses at present, there is intense competition for the available time, with different subject areas vying for inclusion, while there is a clear requirement not to overload students. Both the projects discussed in this paper were only completed by vitrue of the fact that the students involved spent substantially more time working on them than was allocated in the course structure, (approximately 140 hours for the shaft encoder project and 70 for the binary to bcd converter). Such levels of effort on a minor project are difficult to justify, particularly as they may be predominantly involved with becoming familiar with a particular set of ECAD tools, as opposed to the undoubtedly more worthwhile objective of developing sound approaches to integrated circuit design. The precise means by which the 'tool learning time' may be minimised and 'design learning time' maximised, while adhering to a reasonable project hours budget is still a subject of consideration at Aberystwyth. It may be necessary to move away from Silvar-Lisco software for undergraduates and adopt, instead, packages which are less flexible but more quickly understood. In this context, the difficulty of academic staff finding time to become familiar with a wide range of software packages in order to make reasonable recommendations should not be underestimated. Indeed, the programme has been found to be very expensive, though rewarding, in staff time.

Related problems are the burden of providing for the capital costs of workstation replacement at frequent intervals, the cost of hardware and software maintenance and the actual fabrication costs. With regard to the latter, the decision by the Department of Trade and Industry in the UK to pay 50% of educational fabrication costs has certainly been welcomed.

6 CONCLUSION

As a result of the ECAD initiative in the UK, students at the University College of Wales, Aberystwyth have been able to design and have fabricated Texas TAHC gate arrays. The experience has been assessed as worthwhile by both staff and students, and while the lessons learnt suggest some modifications in procedure may be required, there is no doubt that ECAD work is seen as a substantial step forward in the education of professional electronic engineers.

ACKNOWLEDGEMENTS

A number of people have helped to establish the ECAD facility at Aberystwyth, particularly G. Squires and G. Harrold, and their assistance is gratefully acknowledged.

REFERENCES

[1]   *University Working Party on Microelectronics Fabrication and CAD, Abridged Second Report*, unpublished (1985)

ABSTRACTS–ENGLISH, FRENCH, GERMAN, SPANISH

The design of CMOS gate arrays as undergraduate student projects
The design and fabrication of TAHC06 and TAHC10 gate arrays has been undertaken by students studying for a degree in Microelectronics and Computing. This paper describes the incorporation of such ECAD work into an undergraduate degree programme and discusses progress by reference to specific project examples.

La conception de réseaux prédiffusés CMOS comme projets d'étudiants
La conception et la fabrication de réseaux prédiffusés TAHC06 et TAHC10 ont été entreprises par des étudiants suivant les cours d'un diplôme en Micro-électronique et Informatique. Cet article décrit l'introduction de tels travaux en conception assistée de circuits électroniques dans le programme des cours et en discute l'avancement par référence à des projets spécifiques.

Der Entwurf von CMOS Gate-Arrays als Projekt für untergraduierte Studenten
Der Entwurf und die Fabrikation von TAHC06 und TAHC10 Gate-Arrays wurde durchgeführt für Studenten, die für eine Graduierung in der Mikroelektronik und Computing studieren. Dieser Beitrag beschreibt die Eingliederung eine solchen ECAD-Arbeit in ein Undergraduate-Examensprogramm und diskutiert den Fortschritt mit Hinweis auf spezielle Projektbeispiele.

El diseño del CMOS gate arrays como proyecto para estudiante pregraduado
El diseño y fabricación del TAHC10 y TAHC06 gate arrays fue emprendido por estudiantes de grado en Microelectrónica y computadores. Este articulo describe la incorporación de trabajos ECAD en programa de no graduado (de grado) y discute los progresos con referencia a ejemplos de proyectos especificos.

DESIGN, FABRICATION AND TESTING OF UNDERGRADUATE GATE ARRAYS

T. I. PRITCHARD, D. TAYLOR and P. HALLAM
Department of Electrical and Electronic Engineering, The Polytechnic,
Huddersfield, England

1 INTRODUCTION

The Department of Electrical and Electronic Engineering at Huddersfield Polytechnic has been committed to the introduction of electronics computer aided design (ECAD) into its teaching programmes since 1980. Since 1985 the Department has been fabricating undergraduate designs via an industrial semicustom prototyping route. During this time more than 25 gate arrays have been processed, featuring over 50 different designs.

The emergence of low-cost prototyping facilities in the early 1980s prompted the Department, prior to the ECAD initiative, to purchase the HILO and BX logic simulators. HILO was selected as an industry standard digital simulator and BX to access the proven route to low cost semicustom silicon via MCE's FALCON service. The ECAD initiative supplemented and enhanced these existing facilities in the form of Apollo workstations and a range of additional industry standard design software. This design capability has recently been extended to 15 Apollo workstations running up to ten different design packages. Further agreement by the DTI to fund 50% of fabrication costs has enabled design, fabrication and evaluation of prototypes to form a significant part of the undergraduate curriculum.

Further essential purchases made by the Department in support of ASIC prototype fabrication are a Tektronik Data Acquisition System (DAS) and a Wentworth prober for prototype evaluation. We believe that it is essential to be able to make accurate statements about silicon quality and design verification if we are to retain the respect of the silicon vendors used.

This paper describes experiences since 1985 in design, fabrication and evaluation of undergraduate prototype semicustom ASICs. The paper also outlines further developments in the field of full custom ASIC fabrication.

2 BACKGROUND TO UNDERGRADUATE ECAD

Students are now introduced to ECAD as early as the second week of their first year. Throughout the remainder of first year, and into second year, a range of industry-standard CAD tools are encountered (HSPICE, BX, HILO3, SL2000, REDCAD etc). At all stages students are encouraged to compare, contrast and criticise all hardware and software encountered. By their third year, students are sufficiently confident and competent in the use of these tools, and are able to tackle real design problems with the aid of these powerful utilities.

During the third year, working in groups of three, each student has the opportunity to experience the design, fabrication and evaluation cycle via multiproject gate arrays. Each group of students considers itself as a design consultancy contracted to produce a prototype ASIC to replace an existing PCB based design. Contract costs and permitted gate and pin counts are laid down at this stage, with cost penalties for violations. These projects, typically, range from 50 gates up to 500 gates in complexity and usually four different designs are merged to form a single layout to be fabricated on each gate array. The timescale for the design exercise is as follows:

October–	Design (8 weeks)
November	15 hours in total
December	Design merging and layout
	4 hours
January	Fabrication and PCB design
March	Prototype evaluation
	3 hours

Individual student performance is assessed from the contribution to the group design, log book record and formal documentation submitted at the end of the exercise.

Design takes place using the MCE proprietary gate array design software on Apollo workstations, and the destination array is the 5 μm CMOS 1440 gate array. The MCE design software offers:

Schematic capture (or netlist).
Hierarchical design (useful in merging).
Simulator and timing verifier.
Graphical waveform display.
Design database analysis (also useful in merging).
Link into HILO (for ATPG and fault simulation).
Graphical menu for ease of use.

Towards the end of the design stage one group of students takes charge of the design merging process and coordinates the layout of the multi-project array.

3 ECAD EXERCISE 1987/88
The third year ECAD exercise run in 1987/88 will now be described.

3.1 *Design*
In the 1987/88 session four multi-project arrays were processed as part of the third year ECAD exercise. Each multi-project array (Fig. 1) features the four following designs.

3.1.1 *Coin totalliser*   For denominations of 1p, 2p, 5p, 10p, 20p, 50p and £1 the chip should total the coinage entered, up to £9.99, until the total is reset.

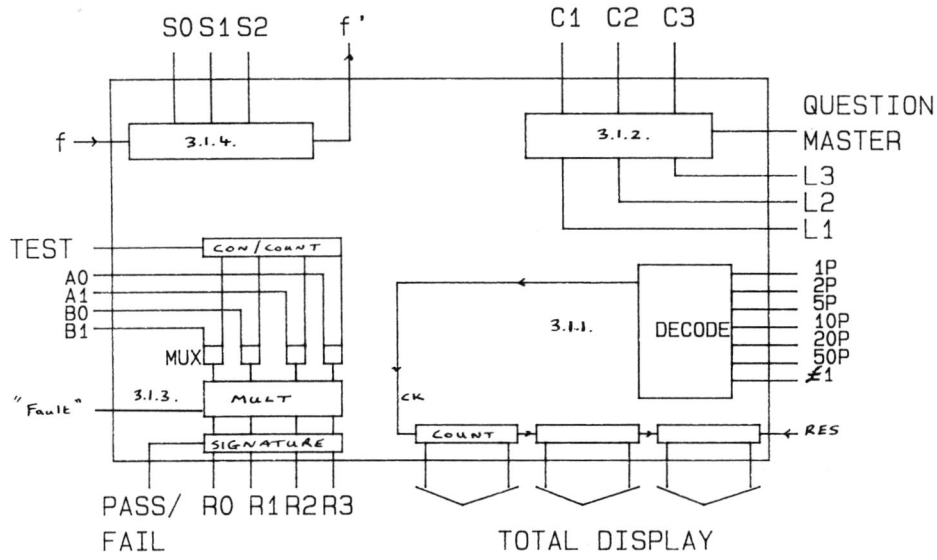

FIG. 1   Multi-project ASIC.

This type of chip could be used in many applications from a vending machine to a moneybox. The design was targetted for 550 gates, and most groups used slightly more. The students found this by far the most difficult of the four designs.

Firstly the students must decide whether to work:

(i)   in BCD on chip for direct interfaces to BCD/7-segment decoders, or
(ii)   in binary for ease of circuit design.

There are also two possible design schemes for the heart of the circuitry:

(a) *An adder based design*	*BCD*	*BINARY*
A bank of adders which adds a number corresponding to the denomination entered to the running total.	Difficult in BCD, must detect inter-digit carry.	Simple design but BIN-BCD for display difficult.
(b) *A counter based design*	*BCD*	*BINARY*
A totallising counter which increments by a count corresponding to the denomination entered. (requires 100 clock cycles) (to update total for £1)	No problem.	Simple design but BIN-BCD for display difficult.

In summary, a BCD totallising counter looks more promising. However, the students must then specify the circuitry which feeds the totalliser. One can either use a single variable modulo counter or a range of counters, perhaps one for each denomination.

These are the type of design decision the students must make, bearing in mind the characterisation of the MCE cell library, the timescales of the exercise, and economical restraints. Marks are awarded for sound logical design decisions, structured design and novel design.

3.1.2   *Quiz panel controller*   For a three-contestant quiz panel there exists a button per contestant and a corresponding light per contestant. The question-master is able to reset all the lights then enable the three contestants buttons. The first contestant to press his button should receive a corresponding illumination and disable the remaining two buttons.

The inherent race condition within such a design makes full use of all the facilities of the simulator during design. The students must determine the best discriminative performance of their design and consider what happens when this is exceeded (do both lights come on?). This is obviously a layout dependent characteristic. This particular design can be achieved in as little as 25 gates but this in no way detracts from its educational value.

3.1.3   *Self testing multiplier (2-bit)*   This design was conceived in order to illustrate self-test enhancement of a combinational logic block. As such the design of the multiplier, within that block, was quite trivial. The self test enhancement consists of a counter multiplexed onto the input of the multiplier and a signature register on the output. As with most such schemes the most difficult part of the design is the test control logic. The control block must latch the 'TEST' signal, clear the counter and signature register, multiplex the counter to the input of the multiplier, count through all 16 input combinations, then stop and return the main circuit into normal mode. The resulting signature is compared with a simulated signature on chip and a single bit pass/fail signal indicates a good or faulty chip. One additional line is used to simulate a fault in the multiplier by tying a line to logic '0', so that the self test will fail. Students are encouraged to assess the self test overheads involved, evaluate shortcomings and the use of such a scheme in larger circuits.

3.1.4   *Selectable frequency divider*   Perhaps the simplest of the four designs, this design takes a particular clock frequency f and from it generates $f, f/2, f/4,$ $f/8, f/16, f/32, f/64$ or $f/128$, depending on the state of three control signals. This is obviously a straightforward clock division and decoding problem and results in around 100 gates. Students carrying out this design finished first and therefore undertook control of the merging and layout phase for the remainder.

3.2   *Fabrication*

In order to simplify the merging phase it is stipulated that each design should be contained in a single module (subcircuit) at the top level of hierarchy. Each of the four designs, in four modules, can then be copied independently into a master group database. Obviously, at this stage, the total merged capacity of the four designs must not exceed the capacity of the destination array, in our case, 1440 gates and 64 I/O pads. If pin limits are exceeded then inputs can be commonned and/or outputs multiplexed between different designs, although this does complicate the merging and layout process. If gate capacities are exceeded then one can easily copy the modules into a larger 3 µm array database and proceed from there.

A manual layout of the merged database is performed, in preference to an automatic one, so that students become familiar with the characteristics of the gate array background. The array is usually laid out as four separate quadrants so that it appears as four separate designs. Within the four quadrants the utilisation of array area is therefore highly efficient. This fine control of the layout process not only cuts the cost of prototyping by almost 50% but also produces a superior product. Students in later years, when utilising the whole of an array, have been known to use 1430 of the available gates on a 1440 gate array.

3.3 *Evaluation*

Whilst the prototypes are being fabricated, PCB test boards are designed and prototype evaluation programmes written for the DAS. On return from the manufacturer, the prototypes are installed on the DAS and the evaluation programmes run. The test programmes are designed to test functionality of the designs and illustrate limitations.

As a final demonstration, the prototypes are installed in the evaluation PCBs and investigated further. The only problems we have experienced have been due to designer error and have been minor (inverted polarity reset lines etc). Extensive tests on selected parts, using the test facilities of HILO3 and the DAS, have shown that usually we are returned 8 perfect parts out of 10. Of the faulty parts, approximately one half have a very minor fault (single output or input stage blown), and the other half a more serious fault.

3.4 *Assessment*

The writing of the final report is a process of formal design documentation, but more importantly the students are expected to bring out the important engineering aspects of the design and prototype performance, as well as making clear conclusions on the use of ECAD, the importance of prototype evaluation and the benefits of custom/semicustom ASICs.

4 FINAL YEAR PROJECTS

Students excelling in the third year ECAD exercise usually go on to perform final year projects in a similar area. These projects run throughout the whole of final year, with a timetabled allocation of 120 hours, and count 25% towards the final degree classification. The whole of any size array may be utilised and the projects are usually industrially initiated and take the form of an ASIC feasibility study for replacement of an existing product.

4.1 *Conveyor belt speed monitor*

This design was carried out as an ASIC feasibility study to a specification provided by a company in the Midlands. The existing design is microprocessor based and retails at £650, with a profit margin of around £100.

The front panel and signal interface to the unit are shown in Fig. 2 and the specification is as follows.

A roller driven by the conveyor is fitted with a sensor, which generates a pulse (or number of pulses) per revolution of the roller. The monitoring unit interrogates this stream of pulses and from it calculates and displays the speed of the conveyor, in ms^{-1}, given the roller diameter. There is also provision for separate user definable 'trip' and 'alarm' signals for underspeed and overspeed, used in controlling the conveyor. In addition the unit provides an analogue signal of between 0.2 V and 4.0 V proportional to conveyor speed, and a ramp waveform for every 10 m of travel.

The aim of the project was to replace three PCBs of microprocessor-based design with a single ASIC, to cut down as far as possible on external circuitry and to make accurate cost and performance comparisons for the replacement.

The ASIC design uses a fast clock (KHz) to time accurately the period of the conveyor derived pulses (Hz). A number of improvements have been made to the original design. The ASIC version displays instantaneous speed as a percentage of rated speed, from 75% up to 115% of rated speed. This was thought to be more meaningful to a user, and, in addition, is independent of the diameter of the roller. The installer simply 'tunes' the unit via a potentiometer to read 100% when running at rated speed. Trip settings are entered as percentages via banks of rotary BCD switches. The analogue part of the specification was realised in external circuitry driven from the ASIC. The next software release from MCE will allow these analogue elements (D/A convertor, operational amplifiers etc) to be created on chip and cut costs further. The final ASIC was around 1100 gates and fits into a 48 pin package. The prototype ASIC was evaluated on the DAS and investigated in a PCB test board. The ASIC was found to perform all of the desired functions.

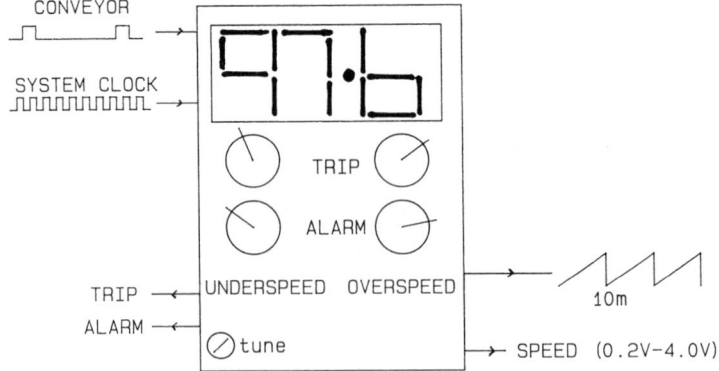

FIG. 2   Front panel of conveyor belt speed monitor.

The student obtained accurate cost breakdowns for the microprocessor based unit and made good accurate comparisons against the prototype ASIC and proposed production ASIC versions. Savings in component and PCB manufacture as well as assembly and test were evaluated. The resulting ASIC version was also a fraction of the size, and therefore, weight of the existing unit and reduced packaging costs could also be predicted. Overall, it was estimated that the ASIC unit could undercut the existing microprocessor based unit by up to 50%, with comparatively low development costs.

4.2 7-day, 24-hour programmable timer sequence

This particular student chose to design a low cost, ASIC-based 7-day, 24-hour programmable timer for use in domestic/industrial heating/lighting applications. The ASIC is single channel, but with capability for expansion, and provides up to 96 'switchings' per day over seven days (672 switchings per week). All the switching information is stored in a low-cost external ROM/RAM which the ASIC accesses whilst sequencing the output channel. Either a sequence ROM can be configured elsewhere, or a RAM can be used since there is facility to continually update, via the ASIC, the sequencing data stored.

Facilities are provided for ease of testing the unit, driving the output channel manually, scanning through the sequencing data at high speed and clearing the sequencing RAM completely. The ASIC power supply is derived from the mains, as is the 50Hz reference clock. The ASIC directly drives an LCD display and needs no external support circuitry. The front panel is shown in Fig. 3.

Typical manufacturers of these units produce between 50,000 and 100,000 units per year. In these production quantities the ASIC could be manufactured for around £2, for an initial development investment of around £20,000, undercutting microprocessor-based products, whilst also providing other ASIC related benefits.

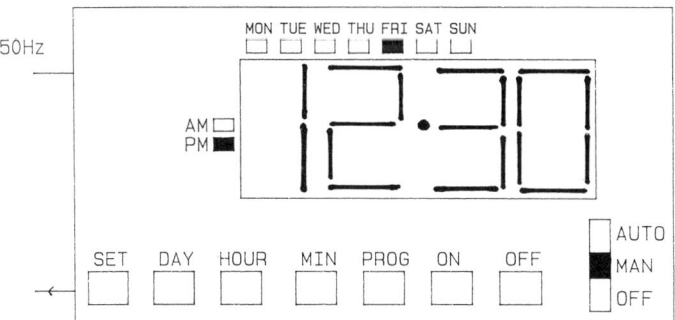

FIG. 3   Front panel and signal interface of 7-day 24-hour programmable timer.

5   FUTURE PLANS

The undergraduate ECAD work described above will continue, and probably expand, over the next few years. Our collaborative projects have been enthusiastically received by industry and we hope to continue and develop these further.

Better analogue and mixed mode design software, coupled with the DTI 50% fabrication funding, has made standard cell and full custom fabrication feasible. It is envisaged that final year undergraduate projects and research projects will utilise these routes in the very near future.

Students currently passing through third and fourth year have not met ECAD in any detail until late in their course. Consequently some time must be set aside for familiarisation with ECAD hardware and software. Students entering the system more recently are introduced to all aspects of ECAD from a very early stage and will be well prepared for more demanding design exercises and projects in third and fourth year.

6   CONCLUSIONS

It is now a requirement by industry that graduates be well versed in many aspects of semi custom and full custom design, fabrication and testing. The ECAD initiative has provided a range of industry standard CAD tools with which to realise this. Industry-related projects also help to promote and develop useful links between educational establishments and manufacturing companies.

Accurate costing of projects provides students with the necessary commercial awareness and entrepreneurial skills required by today's industries.

In this way it is hoped that the ECAD work described above reflects the current design, fabrication and test trends in industry. Students can design large complex ASICs quickly, efficiently and cheaply. Design verification before and after production serves to illustrate the many pitfalls of the design process and limitations of hardware, software and fabrication.

The class exercises encourage a good degree of cooperation and communication both within the design groups and between groups.

All ECAD work has been successful in terms of producing working prototypes, and in motivating and educating undergraduate students. Established and reliable prototyping services provide access to good quality low cost silicon with a guaranteed delivery date.

Most of all, the students find the exercides enjoyable, interesting and relevant.

ABSTRACTS–ENGLISH, FRENCH, GERMAN, SPANISH

Design fabrication and testing of undergraduate gata arrays

The paper describes undergraduate design, fabrication via an industrial prototyping facility, and testing, of CMOS gate arrays. The majority of design projects are initiated by industry, and serve as an ASIC product feasibility study. The technical difficulties and educational benefits of the projects will be discussed, and future developments outlined.

Conception, fabrication et essais de réseaux prédiffusés par des étudiants
Cet article décrit la conception par des étudiants, la fabrication par une entreprise industrielle de prototypes et les essais de réseaux prédiffusés en technologie CMOS. La majorité des projets est à l'initiative de l'industrie et ils sont utilisés dans une étude de faisabilité de produits ASIC (circuits intégrés spécifiques à une application). Les difficultés techniques et les avantages au point de vue éducatif de ces projets sont discutés et les développements futurs esquissés.

Entwurf, Fertigung und Testung von Gate-Arrays in der Ausbildung noch nicht graduierter Studenten
Der Artikel beschreibt für die Ausbildung noch nicht graduierter Studenten Entwurf, Herstellung über eine industrielle Musterfertigung und Testung von CMOS-Gate-Arrays. Die Mehrzahl der Entwurfsprojekte wurde durch die Industrie initiiert und dient als ASIC Produkt-Realisierunsstudie. Die technischen Schwierigkeiten und Ausbildungserfolge der Projekte werden diskutiert und zukünftige Entwicklungen aufgezeigt.

Diseño, fabricación y comprobación de gate-array para pregraduados
Este articulo describe el diseño y la fabricación por medio de la facilidad de un prototipo industrial y la comprobación de gate-arrays CMOS para pregraduados. La mayoria de los proyectos de diseño se emprenden por la industria, y sirven como un estudio de viabilidad de un producto ASIC. Se discuten las dificultades técnicas y los beneficios educativos de los proyectos, y se esbozan futuros desarrollos.

PRACTICAL CHIP DESIGN FOR UNDERGRADUATE COMPUTER SCIENTISTS‡

PETER ROBINSON
Computer Laboratory, University of Cambridge, England

INTRODUCTION

The Computer Science Tripos at Cambridge is a two year course taken by undergraduates after one year's study of a subject such as Natural Sciences, Mathematics or Engineering*. The course includes lectures on digital electronics and computer design but the associated practical work has, until this year, been limited to 10 two-hour sessions in a hardware laboratory where a number of fixed experiments are undertaken. These typically involve the construction of a small circuit from SSI components on a breadboard and investigating its behaviour using an oscilloscope. However, recent developments in computing — notably better CAD tools and cheaper fabrication for integrated circuit designs — have made it possible to introduce a course in integrated circuit design with more creative practical work. Just as a course about a particular programming language might include a practical exercise, so it is now reasonable to include a practical exercise in chip design in a course on digital electronics.

The students were introduced to the Qudos *QuickChip* system[1] in 8 one-hour lectures and 8 two-hour practical sessions. A further 20 or 30 hours were then spent by each student in thinking of a problem and taking it through all the various stages of design to implement it as an integrated circuit. These were manufactured using the Qudos prototyping facility and 5 samples of each design were returned to the students for testing. Almost all the designs were logically correct and the production yield was over 85%.

THE QUDOS QUICKCHIP SYSTEM

QuickChip is a commercial development of research work previously undertaken in the Computer Laboratory at Cambridge[2] and provides a coherent set of tools to design semi-custom integrated circuits. The basic approach is one of correctness by constrained construction. The structure of the circuit is specified using a textual hardware description language (HDL) which is in many ways

‡This article was first published in the British Computer Society's *Computer Bulletin*, Vol. 4, Pt. 3 (Sept., 1988) and is reprinted here by kind permission of the Editor.
*This has now been superseded by a three-year Computer Science Tripos. The course in practical ECAD is taught in the second year of the new tripos.

similar to a programming language. This is the master description of the circuit, and the layout is constrained to implement it precisely. Some automatic tools are available to work out a layout for the chip from the HDL; otherwise the layout can be edited manually, but will be checked for correctness against the HDL in either case.

After thinking of a problem and sketching a suitable circuit, the circuit is transcribed into HDL. This can be performed using a schematic editor, but for this course the students wrote the HDL directly. The HDL is then tested by simulation. The designer prepares a stimulus file containing a sequence of typical input signals and these are injected into the circuit. The resulting changes within the circuit and on its output pins are displayed on a graphics terminal rather like an oscilloscope display. The simulator also produces statistics to show how effectively the test patterns have exercised the circuit and, therefore, how confident the student can be of the design. (This is, in fact, a fault simulation of the circuit, required by Qudos to calibrate testing after fabrication.)

The circuit is then laid out on an integrated circuit. The designer firstly works out a floor plan of the chip, assigning large blocks of the circuit to areas on the chip. These blocks are then elaborated in detail down to the elementary gates provided in the Qudos standard library. QuickChip includes various tools to place and route the components on the chip, and these can be further edited by hand. In either case, a hierarchical checker is used to verify that the chip designed corresponds exactly to the original HDL.

Finally, the parasitic capacitances induced by the layout can be extracted and the circuit re-simulated to see how it will perform when analogue behaviour is taken into consideration. At the same time, the simulator can generate a test pattern for wafer probing after fabrication.

Qudos' Silicon Bus service is a fabrication service for prototype semi-custom integrated circuits by commitment of one or two layers of metal using direct writing with electron beam equipment[3]. The service currently supports both Ferranti C series bipolar uncommitted logic arrays and Texas Instruments 2μ TAAC and 3μ TAHC series CMOS gate arrays[4], the latter being used for this exercise. For educational designs, testing by Qudos is limited to extraction of parameters from control structures on each wafer and checking that individual circuits draw roughly the right current; in practice, very high yields were obtained. Turnaround can be as long as two months, depending on the arrival of sufficient educational designs to fill a wafer. Needless to say, anyone prepared to pay the commercial rate for fabrication can expect turnaround in a few days and full testing.

THE COURSE
The course is intended to fulfill two educational objectives: introducing the students to the principles of computer aided design for integrated circuits and providing practical experience of hardware design. It is increasingly common in computer science for hardware to be regarded simply as a way of speeding up

algorithms, and computer scientists need to be able to use hardware just as easily as they would use a different programming language or processor.

QuickChip exhibits all the properties found in the best CAD systems while retaining a coherent, integrated approach that makes it simple to learn and use. The elementary tutorial guide prepared for this course (and now used by Qudos) requires only 100 pages to lead a complete beginner through all the stages of design. Nevertheless, the students are aware of the use of a hierarchical hardware description language, simulation and fault simulation, floor planning and detailed layout, circuit extraction and matching, re-simulation to take account of parasitic capacitances, and a very elementary introduction to the principles of CMOS technology. Given this experience they should then be able to absorb other CAD systems fairly easily.

The 8 one-hour lectures and 8 two-hour practicals take about a quarter of the students' time for half a term, and the 20 or 30 hours' work to complete their own designs is comparable to the time taken on each of the 4 programming exercises set through the year. Overall, only slightly more effort is required than in learning a fairly complicated programming language. Of course some of the more imaginative students spend a great deal of time on their projects and come up with designs that could actually be marketed. Others return to the CAD system in their final year to undertake a more substantial project which is written up as a dissertation and counts for a third of their final marks.

THE PROJECTS

The practical ECAD course was first tried in a year with a particularly small class — just 28 students — which meant that the budget could accommodate a whole TAHC06 chip for each student, with potential for the equivalent of 600 two-input gates. The practical work was optional, nevertheless 21 students submitted individual designs and one pair submitted a joint design; a further design was submitted by a member of staff. $22\frac{1}{2}$ of these 23 designs were logically correct and the yield of the samples was over 85%. (The half design that did not work had an error in the layout that was flagged by the CAD system and ignored by the student!)

The sample chips were tested in the hardware laboratory using the same breadboards and oscilloscopes that had been used for the earlier hardware practicals. A collection of standard SSI chips was available to build driver circuitry and output displays.

The students were given a completely free rein in the choice of circuit, and were encouraged to submit something for fabrication even if it was only a trivial exercise from the earlier practical classes. In the event, five of the designs were essentially trivial, being fairly small, purely combinational logic (decoders, multiplexors and so on). Five more of the student designs and the staff design were reasonably complicated, using about 200 gates including a fair amount of internal state. These included circuits for a simple 4-bit ALU, a cascadable 4-bit multiplier, a bi-directional Grey-code generator and a couple of games. The remaining designs were of intermediate complexity, involving a few bits of internal state and some surrounding combinational circuitry. Traffic light controllers, BCD counters and electronic dice were all popular. Only five of the designs actually used more than half of a chip, which suggests that it would be reasonable in general to put two projects on each chip, allowing a full chip to anyone who wanted to pursue a particularly ambitious design.

On reflection, internal state should have been an important consideration. One of the educational objectives of a course in chip design is explaining the notion of testability and, in the absence of internal state, that becomes a null issue. Preparing the stimulus files for simulation had given some indication of how the chips would be tested, but characterising their performance at high speed was not easy with simple oscilloscopes. It might have been worth mentioning self-testing explicitly.

COSTS AND BENEFITS

The costs associated with the course can be divided into two components: the cost of providing the equipment on which the CAD software runs and the cost of fabrication for the practical designs. It is significant that QuickChip does not require a special-purpose CAD workstation and that the fabrication costs are compatible with each student working on an individual design.

The QuickChip system runs on a wide variety of computers and terminals; this exercise used a network of microVAXes under the Ultrix operating system connected to BBC computers with mice serving as graphical terminals. Teaching for the Computer Science Tripos at Cambridge already uses this equipment for practical work, so it was convenient to run QuickChip directly on it. In this configuration it was possible to arrange that each student had an individual terminal at practical classes; this is particularly valuable, because a great deal is learned by actually being the person to push the buttons. The total cost for a share of the VAX and the communications equipment, together with the terminal, amounts to about £2500 per seat, which compares very favourably with other schemes. Of course, the equipment is totally general-purpose and is used for other practical classes as well; indeed its ability to handle colour graphics and interaction with mice makes it suitable for a wide variety of work.

The total cost of fabrication to academic institutions is below £300 for five samples of a design on the TAHC06 array, with turnaround in under two months. However, the recent government initative to promote electronic CAD in UK universities, polytechnics and colleges (which includes the full Quick-Chip package) has been extended to subsidise half of the cost of fabrication of prototypes. This, together with the use of multi-project chips (or, at least, two-project chips), brings the cost of fabrication down to under £100 per design. At this cost, it is quite reasonable to allow each student to pursue their own individual design.

It does seem important that the designs are actually fabricated and that each student works on an individual design. With a CAD system such as Quick-Chip, the fabrication is something of a formality: as the results demonstrated, if the system says that a chip will work, then it will work (and, unfortunately, if the system says that it will not work, then it will not work). Nevertheless, there is a world of difference between seeing a circuit simulated and connecting power to the real thing. Many practising engineers believe, in theory, that application-specific integrated circuits are feasible but somehow lack the conviction to use them in projects; the students who took this course have no doubts about the technology or their own abilities. There are also important practical lessons about debouncing switches and using LEDs of the right polarity that are unlikely to be learned from a simulator.

Working on an individual project is also important in this context. If a pair of students work together, there is always a temptation for one to push the buttons and the other simply to watch, thereby failing to find the confidence that comes from successful practical work. Learning to work in a team is an important skill, but it is a separate skill and is best taught and learned separately.

One consequence of this approach is a significant demand for seats in practical classes. Most of the teaching about the CAD system was undertaken at the practicals. The lectures simply served as an introduction to the system and as an opportunity to fill in some of the background information about the principles of CMOS technology, VLSI design and electronic CAD. A typical class of 60 students would be divided into two groups, each requiring 30 seats for practicals. The cost of this would be prohibitive if special-purpose CAD workstations had to be used.

CONCLUSIONS

Perhaps the most significant comment on the course was the enthusiasm which it inspired amongst the students. Even at Cambridge it is rare to find students turning up half an hour early for practical sessions and returning to complete a voluntary project after the end-of-year exams. However, one or two did express concern at the cost of the exercise and, indeed, at £300 per student it was not trivial. Even so, it does bear comparison with other laboratory subjects and the reduction of costs to under £100 per student makes it quite acceptable.

Of course, chip design has long been part of specialist courses, usually at the postgraduate level and with an emphasis on engineering aspects. However, it is still unusual for Computer Science graduates, who will spend most of their professional lives working with software, to have practical experience of hardware design in general and integrated circuit design in particular. They need to recognise hardware as just another tool to solve problems in Computer Science, alongside the various high-level languages and assembly codes. The recent simplification of chip design techniques, as embodied in packages such as QuickChip, means that it is now possible to teach hardware design at the level of computer systems. The fall in the cost of fabrication of prototype circuits means that a course in Computer Science can realistically supplement this teaching with practical work involving the design and manufacture of integrated circuits.

A further effect of this course has been a change in attitude of research students working on hardware projects in the Laboratory. Several attended the practical ECAD course and are now using QuickChip to design integrated circuits that will be used in their own experimental work.

REFERENCES

[1] Hopper, A., *QuickChip and the silicon bus*, 6th International Conference on custom and semi-custom ICs (1986)

[2] Robinson, P. and Dion, J., *Design aids for uncommitted logic arrays*, 2nd International Conference on semi-custom ICs (1982)

[3] Shelley, T., *Low-cost machine logic revolutionised*, Eureka innovative engineering design, (December 1985)

[4] *TAHC series CMOS gate array reference manual*, Texas Instruments (1983)

ABSTRACTS–ENGLISH, FRENCH, GERMAN, SPANISH

Practical chip design for undergraduate computer scientists
A new course in practical electronic CAD has recently been introduced to the Computer Science Tripos in Cambridge. This note describes the way that the course was organised, discusses the sort of chip that was designed and concludes with an assessment of the educational benefits.

Conception pratique de puces pour des étudiants en sciences informatiques
Un cours nouveau en CAO de circuits électroniques a été récemment introduit dans le Computer Science Tripos à Cambridge. Cette note décrit la façon dont le cours a été organisé, discute du type de puces qui a été conçue et conclut par une évaluation des bénéfices éducatifs.

Praktischer Chip-Entwurf für Studenten der Computerwissenschaft
Kürzlich ist in Cambridge ein neuer Kursus für praktischen computergestützten Elektronikentwurf zum Computerwissenschaft-Tripos (letztes Examen für den honours degree) eingeführt worden. Diese Notiz beschreibt die Weise, in der dieser Kursus eingerichtet wurde, bespricht die entworfene Chip-Art und endet mit einer Bewertung der Erziehungsvorteile.

Diseño práctico de chips para científicos de pregrado en computadores
Se ha puesto en marcha recientemente un nuevo curso en el uso del ECAD en el Computer Science Tripos de Cambridge. Esta nota describe cómo se ha organizado el curso, discute el tipo de chip diseñado, y concluye con una valoración de los beneficios docentes obtenidos.

COMMERCIAL ECAD SOFTWARE IN AN M.Sc. GATE ARRAY DESIGN EXERCISE

J. MACKLE, S. H. S. MAGILL, D. H. CAMPBELL and
J. H. MONTGOMERY
Department of Electrical and Electronic Engineering, The Queen's University of Belfast, N. Ireland

1   INTRODUCTION

Commercially, full-custom (and to a lesser extent, standard cell) application specific integrated circuit (ASIC) designs are only justified where the high development costs can be amortised by high volume production exploiting their superior chip area efficiency or where the costs are of secondary importance to performance in specialist applications such as military or avionics. Consequently, with their low volume cost effectiveness, gate arrays have captured an increasingly wider share of the ASIC market.

Educationally, as an exercise in IC design, the choice is less clear and is subject to a variety of trade-offs. A standard cell approach affords the opportunity to combine both analogue and digital blocks, but for purely digital designs it offers little educational advantage over a gate array approach while being considerably more expensive. A full custom approach is educationally attractive as it has the potential to provide experience in a full range of the necessary skills. In practice however, as time is usually a restrictive factor, the device level experience gained can be at the expense of other skills. This is aggravated by the fact that full custom/standard cell turnaround times are typically 2 to 3 times longer than those for gate arrays. Because of their fast turnaround times and highly automated design procedures, gate arrays facilitate the design of more complex digital circuits in a given time. This can provide experience in areas which are of critical importance in complex circuit design, such as IC testing and 'design for testability'.

Ultimately, the choice will depend on the balancing of the respective trade-offs with localised factors, such as: time and resources available; student/staff ratios; and the priorities and objectives of the exercise itself. From the foregoing, a gate array approach was considered the most suitable design vehicle for the University's M.Sc. in Electronics. Students keen to add full custom experience can still do so, on a more selective basis, in their major project.

This paper describes how a variety of commercial ECAD software is incorporated into the gate array design exercise and details the timetable and projects associated with the academic year 1987–88.

2   THE QUEEN'S UNIVERSITY OF BELFAST M.Sc. IN ELECTRONICS

The Queen's University of Belfast postgraduate M.Sc. course in Electronics is designed to cover a range of modern electronics topics while affording the opportunity to specialise in several key areas. Integrated circuit design is one such key area and this option became one of the SERC supported IC design courses in 1981.

The M.Sc. course commences in October and normally takes 12 months. It consists of two terms of lecture material and an extended project which is completed during the Summer. Students are assessed separately in written examinations on the lecture material (held in April) and in the project work including presentation of a dissertation.

The course may also be taken in modular form, by part-time students over a two or three year period, students being required to take at least two topics per year and to meet the project requirement.

The lecture course consists of six topics chosen from the following list:

Computer-aided analysis and design; Custom integrated circuit design; Digital system design; Microprocessor control and software engineering; Computer systems; Microelectronic technology; Electronics of solid state devices; Information and communication theory; Microwave and optical communications technology; Analysis and synthesis of passive networks.

The first three topics form the core subjects of a specialisation in IC design and are complemented by a gate array design exercise, included as part requirement of the custom integrated circuit design option.

3   THE DESIGN TOOLS

The gate array currently used is the 5 micron FALCON array from Micro Circuit Engineering (MCE). This is a single layer metal, silicon gate CMOS array with 1440 gates (nominal) and 64 pads. As part of their low cost prototyping service (available through the Electronics Computer Aided Design (ECAD) initiative), MCE provide 10 prototype devices in either 40 or 64 pin ceramic DIL packages, from designs submitted in netlist or layout form. A test circuit is included in each die and is tested at the wafer stage to ensure correct processing. Apart from this, the prototype devices are returned untested.

From the CAD software currently available through the ECAD initiative, the route from 'paper design' to MCE-format netlist can be effected in two distinct ways. The first is to use MCE's own integrated CAD package 'BX DESIGN' which includes schematic capture, simulation and timing verification. The second, is to follow the route described in Fig. 1. Here, Silvar-Lisco's SDS (Structured Design System) suite is used for schematic capture and netlist generation and Genrad's HILO is used for simulation, fault analysis and timing verification. The conversion programmes MCEWAVE and MCECCT, and the HILO models and CASS symbols for the MCE primitives, have been provided by SERC. MCEWAVE produces an MCE-format simulation waveform file directly from the HILO simulation results and MCECCT converts the SDL (Structured Design Language) netlist, produced by IGEN, into MCE format.

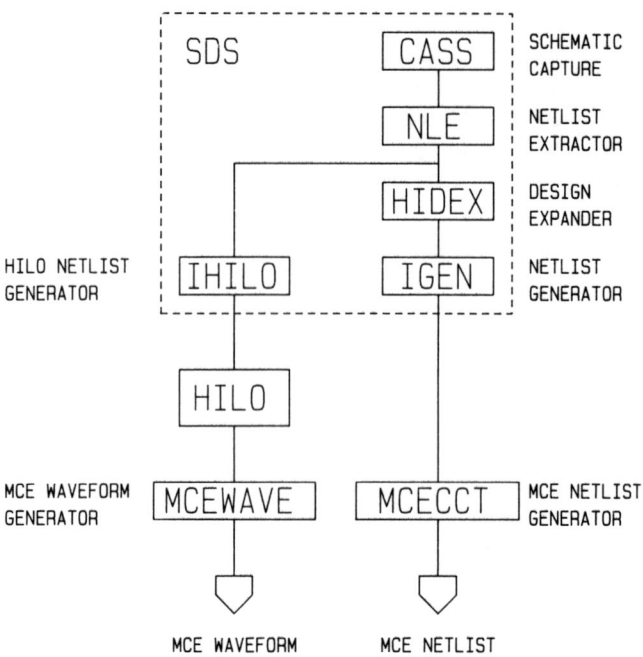

FIG. 1   The design route.

BX DESIGN has been designed to be both simple to understand and easy to use, however, educationally the second route was considered more beneficial as students would gain a wider experience of more sophisticated commercial ECAD software, while still designing on a commercially viable and available gate array.

4   THE GATE ARRAY DESIGN EXERCISE

For the academic year 1987–88, 16 students chose to specialise in IC design. A 3-node Apollo network, consisting of two disked, colour DN3000s and a single diskless, monochrome DN300, was available exclusively for the design exercise. The DN3000s were bookable for a maximum of 2 hours per day per student between the hours 9.00am–9.00pm and, outside this time, were available on a first come–first served' basis. The DN300 was available on a first come first served basis at all times, although it only proved popular as deadlines approached.

With the written examinations being held in April, the design exercise was restricted to 18 weeks duration, beginning in mid-October and ending in mid-March. The students followed the design route described previously and illustrated in Fig. 1, and were strongly advised to adhere to the recommended timetable (Table 1).

The students were first introduced to the AEGIS operating system and the SDS suite and familiarised themselves with these by means of a simple example

TABLE 1   *Timetable for design exercise.*

WEEK	ACTION
1-2	Familiarisation with design system
3-4	Verification and partitioning of proposed specification
5-8	Schematic entry
9-10	Familiarisation with HILO
11-17	Circuit validation and test pattern generation
18	Data conversion to MCE format

circuit (a divide-by-50 ripple counter). Design projects were assigned either singly or in pairs and, as this was not to be a digital design exercise, each project was supplied as a detailed functional specification, together with a basic (and not always error free) solution. Each student was required to check and minimise the given logic, introduce hierarchy where expedient, and to modify/partition the circuit to improve testability if required. The complexity of the projects ensured that these steps were seen to be both necessary and worthwhile.

Table 2 lists the projects for the 1987–88 course together with their approx-

TABLE 2   *1987–88 projects.*

	PROJECT DESCRIPTION	GATE COUNT	NUMBER OF STUDENTS
CHIP 1	8-contestant quiz referee	150	1
	2-digit, presettable down timer with alarm and 6-range clock generator	500	2
CHIP 2	3-digit LCD, presettable, up/down photographic timer with alarm	700	2
	6-range crystal clock generator	350	1
CHIP 3	6-digit LCD, MSF Rugby clock decoder	850	1
	24-to-12 hour convertor	150	1
CHIP 4	Decimal keyboard decoder with 16-bit tri-state output and 4-digit multiplexed display	650	2
	Thermal inertia heater controller, percentage power set via BCD input or keyboard decoder	200	1
CHIP 5	16-bit pseudo-random number generator	200	1
	4-digit BCD-to-binary convertor	350	1
	16-bit binary-to-hex decoder with 4-digit multiplexed display	300	1
CHIP 6	2-digit up/down timer/controller with alarm and 6-range clock generator	650	2

imate gate count, and illustrates how the projects are functionally specific and complete, either singly or in combination, rather than just generic counters or decoders. This has proved to be a great motivating factor, as thorough design work is rewarded with a useful and impressive 'box of tricks'.

The remainder of the first term was occupied with schematic entry of the circuit designs. After a brief introduction to HILO, the second term was almost fully occupied with simulation, circuit validation, timing verification and test pattern generation. In most cases this also involved returning to schematic entry to correct errors uncovered by the simulations. The final design stage for the student was to convert their circuit netlist and simulation waveforms into MCE format. A brief comparison was made with BX DESIGN.

Once simulation was completed successfully, staff combined the functionally related projects, resulting in six distinct chip designs. This was effected using the same route as the students and the respective simulation waveform files were similarly combined, to produce a complete test for each design, in MCE format.

In order to have as much control over the designs as possible and to reduce costs, five of the designs were layed out using MCE's 'BX LAYOUT' software. These were re-checked with post-layout simulation and submitted to MCE for fabrication, along with the sixth design which remained as a netlist. Fabrication took four weeks and the ICs were tested using a Tektronix DAS9100 SERIES digital analysis system. The original simulation waveforms and results were used for this test but at present, the programming of the pattern generator, and the interpreting of the results from the logic analyser, both have to be performed manually. All the designs were successful and a yield in the range 50%–90% was returned.

5   CONCLUSIONS

The gate array design exercise described, has been in operation for two years and to date, 30 students have successfully completed this route, producing 14 custom ICs, in all.

The fabrication of the designs has proved to be a considerable motivating factor and, in the students' minds, lifted the exercise above one of mere 'academic' interest. This is particularly so, with the use of 'application specific' projects, whereby demonstrations of previous student designs provided an added incentive.

Due to the time constraints imposed by the necessary structure of the M.Sc. course, the layout and chip testing stages are not yet a formal requirement of the design exercise. However, the majority of students took a keen and active interest in both these important stages and it is hoped to incorporate these formally in the near future.

ACKNOWLEDGEMENTS
The authors would like to express their gratitude to the SERC for providing the financial support and resources necessary to make the IC design option on

the M.Sc. course possible and to MCE for their co-operation in producing the gate arrays. The authors are also indebted to those colleagues at other U.K. universities who have devoted their time and energy in making the Higher Education ECAD Initiative so successful.

ABSTRACTS–ENGLISH, FRENCH, GERMAN, SPANISH

Commercial ECAD software in a MSc gate array design exercise
By incorporating a variety of commercial ECAD software into a gate array design exercise, students on a one-year MSc in Electronics are able to gain both practical and diverse experience in industrially relevant semi-custom design techniques.

Logiciel commercial ECAD dans un exercice de conception de réseaux prédiffusés au niveau maîtrise
En introduisant divers logiciels commerciaux ECAD dans un exercice de conception de réseaux prédiffusés, des étudiants d'un cours de Maîtrise en Electronique en un an obtiennent une expérience pratique et diversifiée dans les techniques industrielles de conception de circuits prétraités.

Kommerzielle ECAD Software in einem M.Sc.-Gate-Array-Entwurfskurs
Durch Einbeziehung einer Vielfalt von kommerzieller ECAD-Software in eine Gate-Array-Entwurfsübung, werden Studenten im 1 Jahr-MSc-Kurs in Elektronik befähigt, Praxis und auch diverse Erfahrungen in industrierelevanten Semikunden-Entwurfstechniken zu erwerben.

Software comercial ECAD para un ejercicio de diseño de un gate-arrays en curso de Master
Incorporando la variedad de software comercial ECAD en un ejercicio de diseño de gate arrays, los estudiantes de un curso Master en Electrónica de un año son capaces de conseguir experiencia práctica y variada de tecnicas de diseño semi-custom industrialmente importantes.

ASIC TECHNOLOGY APPRECIATION: A 'HANDS-ON' APPROACH

H. L. JONES
Department of Electrical Engineering, Polytechnic of Wales

1 INTRODUCTION

Arising from the well-publicised Electronics Computer Aided Design (ECAD) initiative, all higher education establishments within the U.K. were offered commercial-quality computer aided design software packages, primarily for full and semi-custom integrated circuit implementation. Attractive deals were negotiated between the vendors involved and the academic community. With the co-operation of the Department of Trade and Industry, amongst others, five regional co-ordination centres or 'lead sites' were set up to handle software distribution and technical problems, thus relieving the commercial suppliers from the burden of maintaining the low cost and widespread issue of their products. A unique opportunity had thus arisen for those polytechnics and universities that had hitherto not been actively involved in Application Specific Integrated Circuit (ASIC) technology, due to prohibitive costs coupled with lack of expertise. Financial assistance, in the form of grants and subsidies, eased the purchase of state-of-the-art computing equipment which was needed to run some of the ECAD packages.

There is a common fear that if CAD work is formally introduced into a curriculum it will displace some other fundamental lecture material. Whilst recognising this hazard, the interested staff in the Electrical Engineering Department at the Polytechnic of Wales wished to make use of the new CAD tools at the earliest possible date, without having to resort to re-arranging timetables or the technical content of any subject. In the second year of the B.Eng. degree scheme, provision had been made for a group project. In previous years this had been executed as a number of disparate tasks associated with a specific course subject. Students worked on just one of these projects, two hours per week, for the academic year. It was decided that an integrated circuit design group project would fit in with the B.Eng. scheme. Furthermore, to maximise the number of students gaining such a valuable exposure to ASIC design, a single project specification would be tackled by all the groups.

To fulfil the aims of the group project, practical and tangible evidence of the work should be evident at the conclusion of the project, i.e. actual chips should be manufactured. Without any previous terms of reference, the supervisory staff had no firm guidelines as to the feasibility of co-ordinating such a project, which was felt to be ambitious due to the large numbers of students involved. One package available under the ECAD initiative, 'MINICHIP' by Qudos of

Cambridge, seemed to offer most promise for the proposed exercise. Marketed as an educational design aid it runs on enhanced versions of the BBC micro-computer and, whilst not purporting to be a professional software tool, offers the option of actual custom chip implementation by Qudos themselves.

2 PROJECT SPECIFICATION

As a multi-disciplinary vehicle for reinforcing students' engineering, self-study and management skills, the group project required that the design of a semi custom IC be just one facet of the overall specification. A design target was required that had to be challenging yet realisable. A project proposed by the image processing group at the University of Wales Institute of Science and Technology (UWIST) in Cardiff seemed a suitable candidate.

In image processing, many of the algorithms involve convolution of a digitised picture with a 3×3, or similarly dimensioned, array which is passed over the complete pixel representation in memory to give a new resultant image with enhanced contrast or vertical edges, for example. One such operation is the rank filter, which re-orders pixels centered around a window matrix according to intensity levels. A feature of high resolution image processing is the computationally intensive nature of even the most rudimentary processing task[1]. Thus, real-time image processing can require specialised, fast hardware to perform a particular function that would be too slow if attempted by software.

To implement the rank filter as an item of hardware, a novel approach was proposed whereby the pixel stream from a video camera, a series of 8 bit binary values, would be passed through a 9 pixel window. Thus a nine byte FIFO is

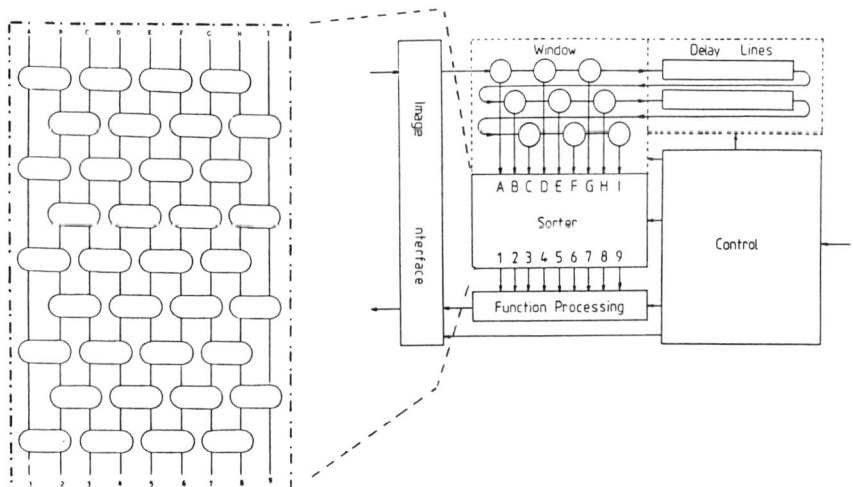

FIG. 1   Rank sorter within image processing. Each oval is one custom IC. The smaller of the two input bytes is routed to the left, the larger to the right.

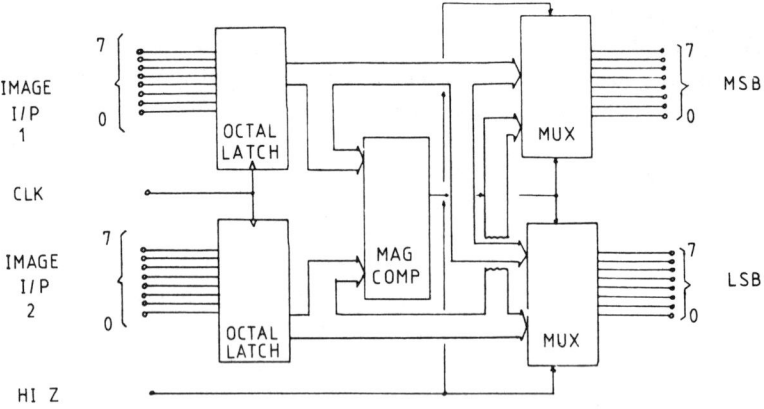

FIG. 2   Functional overview of custom IC.

required with the outputs fed, in parallel, to a rank sorter section. Coefficients within the window element would perform any requisite pre-processing function whilst the sort unit would output intensity regulated video samples (Fig. 1). Many standard components are required to synthesise the rank sorter since it contains 36 identical stages to perform the ordering.

Each block consists of a data entry latch for synchronixed pixel transfers, an 8 bit comparator and an octal multiplexor to route the samples. A diagrammatic representation of each identical sort section is given in Fig. 2. This structure was implemented as a custom chip in a 40 pin DIL package. Provisional design estimates intimated that the implementation would be equivalent to 150 two or three input NOR gates, which was the level of complexity preconceived as suitable for the group project.

All the student groups are expected to delegate and co-ordinate their efforts with minimal supervision after being assigned a specification. Within each group, students can concentrate on areas of individual interest. A project leader must be appointed who controls the overall progress of the work. This was defined as implementing a rank sort unit as a gate array and developing automated test equipment to verify the operation of the ICs produced. A marketing and research element is also a constituent of the project.

3   PROJECT ADMINISTRATION

Sixty- four students, in total, were involved in the ASIC design exercise. A network of BBC Master series computers were installed to run the Qudos suite, comprising one hard disk file server, a print server and eight work stations. Due to laboratory accommodation problems, the class was split into four and a single two-hour session per week allocated to each quarter. This implied splitting the cohort into eight teams of eight people so that two groups could be in the laboratory at any one time, i.e. four terminals per project.

A small number of lecturers acted as advisors for all the project teams instead of assuming responsibility for individual groups. This meant knowledge could be distributed as staff and students alike learned from experience. Preliminary time schedules were issued with seven phases identified. Four weeks were allocated for MINICHIP tutorials and exercises for all students before embarking on the project proper.

For prototype chips, the service offered by Qudos is to supply five parts for £300.00 from a single design submission. Due to budget constraints, each group could not have a unique design processed. Since a common specification was given to all groups it was decided to introduce a competitive element in that funding would be made available for the manufacture of just one design, which would be chosen after a formal presentation by all the groups, who would have an involvement in the selection procedure. As an incentive, the winning team would provide documentation to the other groups who would then adopt this design for the test equipment development, possibly involving reiteration of previous work. Qudos agreed to supply 8 samples of the design, one for each group, for £400.00. Thirty days turn-around-time was quoted provided the so-called 'Silicon Bus' conditions were satisfied.

4   QUDOS 'MINICHIP'

Emulating a full scale semi-custom IC design cycle, the QUDOS Educational CAD suite allows users to implement logic circuitry on the ULA9C gate array, manufactured by Ferranti (now part of Plessey Semiconductors Limited). All the software is written in BBC BASIC and runs on a BBC Model B or Master series computer if fitted with a 65C102 co-processor, either as a network or from floppy disk. Certain limitations must be accepted, but the system does offer students a good insight into ASIC practice.

Custom chips are realised in terms of NOR gates and inverters. Each of the 990 cells inside the ULA9C contains four uncommitted bipolar transistors which are connected together in pre-defined patterns to form valid logic gates. A border of 64 pad driver circuits, or peripheral cells, surround the cell matrix, organised as 9 blocks of 11×10 cells. Unfortunately, due to processing speed and memory size limitations, MINICHIP can only handle up to 300 gates for a single design but personal experience suggests an upper bound of say 200 NOR gates.

From an initial circuit concept to final manufacture using the Qudos system the stages would be:

(i) Draw a circuit diagram in terms of NOR gates, inverters and standard pad connections. Assign node names for all gate outputs.

(ii) Convert this schematic into a net list description or HDL (Hardware Description Language) and run the 'PARSE' program to generate a design data base.

(iii) Define a test pattern for use in logical simulation. This file would specify a series of input pad events or stimuli to check functionality. Run 'STIM' to create a test data file.

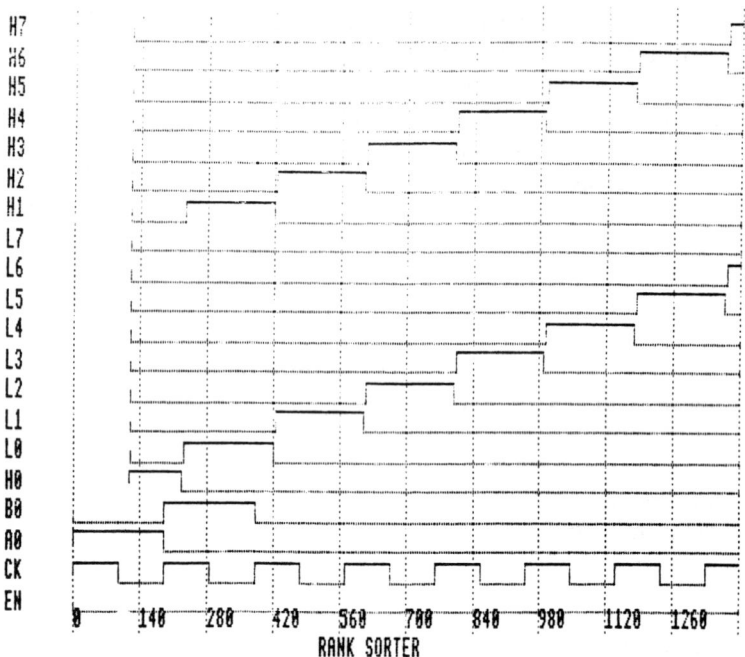

FIG. 3   MINICHIP simulation output.

(iv) Use 'SIM' to simulate the operation of the circuit and hence validate conformance to specification. This program uses the previously compiled data files to produce a logic analyser type trace on the screen (Fig. 3). Debugging facilities are provided to assist in fault location.

(v) Perform the metallisation of the circuit using 'LED'. This is an interactive, manual process using a graphical image of sections of the chip. Plots can obtained from a dot matrix printer as shown in Fig. 4. This data is used to create the final devices. Peripheral cells accommodated are TTL input, TTL output and tri-state output.

(vi) Since the interconnection was not automated, a file must be created using 'NED' to cross-reference the circuit database with the gate array cell positions occupied and pads used.

(vii) An initial check is required to establish that correct gates have been fabricated in each cell. A program 'NCHECK' analyses that no metal tracks have omitted and that each cell will work as the gate type declared.

(viii) Run 'LCHECK' to ascertain that the gates have been interconnected properly. Capacitive delays are extracted at this stage which take into account the length of metal tracks and the additional propogation times expected through polysilicon crossunders.

(ix) Re-simulate the design taking the track capacitance effects into account. It is advisable to introduce a time dilation factor to allow for temperature and processing tolerances and repeat the simulation until satisfactory.

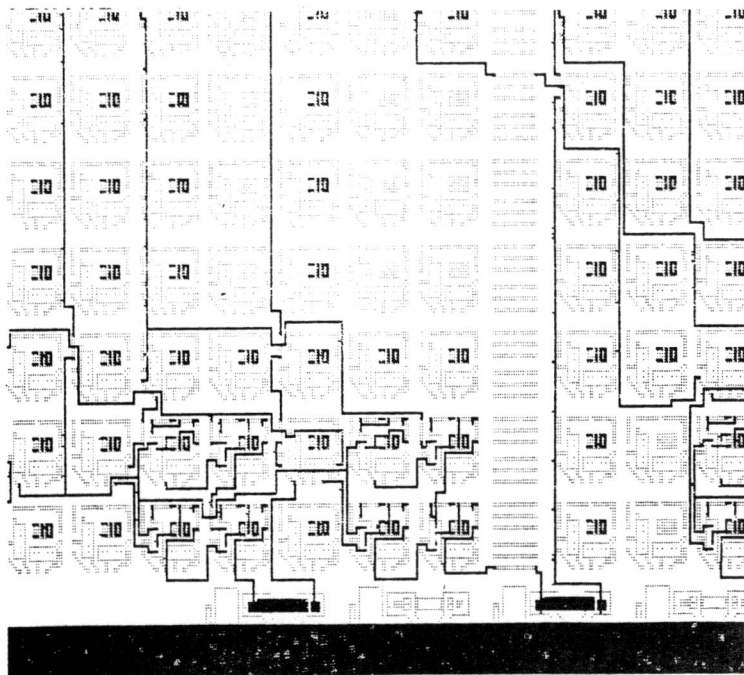

FIG. 4   Graphical plot of area around pad PDF.

At any time in this sequence, the user might have to 'backtrack' and make modifications before proceeding. Being written in an interpreted language, this software is rather slow. Apart from 'LCHECK' which took over two hours to process a completed design (which was approximately 150 gates as forecast), the major drawback was found to be 'No room at…' errors as some of the programs in the suite could quickly exhaust the available memory within the machine. Despite its convenience, the Acorn Econet network is less than ideal as the MINICHIP environment. During checking and metallisation, frequent data transfers are made over the network to the file server. Bottlenecks can occur with one or more stations losing contact with the file server. In the final stages of the design, where several groups were engaged in lengthy checking procedures, it was found necessary to convert some BBC Masters to disk based machines with MINICHIP running from floppy disk.

By the notional cut-off date, three chip designs out of eight were complete. One was selected for submission to Qudos. Following the recommendations of the user manual, a test pattern file was created using 'TVIEW' which accompanied the design data files which were despatched to Qudos at the beginning of March, 1987, for fabrication. Allowing for thirty days turn-around, this meant that the students could concentrate on other aspects of the assignment before being able to utilise the gate arrays, at least three weeks before the group project concluded for examinations. A full-time post-examination period had been scheduled which was intended for collating the technical reports and final debugging of any hardware.

5 APPRAISING THE CHIP DESIGN EXERCISE

Unfortunately, Qudos could not meet the quoted delivery dates. In fairness, the fabrication price and manufacturing times are dependant upon Quodis receiv-ing ten MINICHIP designs for economic processing of a ULA9C wafer. After a lengthy delay in which no other educational designs were submitted, Qudos made at least two abortive efforts to produce the rank sort chip, but to no avail. A faulty batch of wafers was cited as the problem by Qudos, but, what-ever the reason, it meant that the students did not obtain a custom chip and the final project presentations had lost impetus. After their earlier efforts, this was a grave disappointment to all the students; each group had given far more man-hours than timetabled for. There was an abundance of enthusiasm throughout the design phase. After the introductory weeks, up to four people in each group tended to specialise on the gate array design.

As a system, MINICHIP lived up to expectations. Students found it easy to learn, though cumbersome. Certain software problems were located during the project which was a measure, perhaps, of how extensively the system was used by the groups. Qudos were most co-operative in terms of technical advice. Eventually, free of charge, Qudos supplied five sample parts in November 1987. This was six months overdue and all the students involved were undertaking industrial training placements and were no longer available to evaluate the chips. In the absence of the students, it was decided to investigate this batch for logical correctness only; parametric measurements i.e. operating speed, were to be ignored.

For expediency, a test circuit was designed which interfaced with the TUBE of a BBC Microcomputer. Two 2MHz 68B21 PIAs were connected to the 6502 data bus and the output pins of these devices were connected to a 40 pin ZIF socket for the DUT. Thus, a software-based approach was used which, essen-tially, configured the PIA ports accordingly, injected test patterns at the sort chip inputs, clocked the data and monitored the output latches for correct ordering and valid bit patterns. Screen-based error messages provided limited diagnostic information where the rank sorter IC did not behave as predicted. Despite all the chips having similar power consumption figures, only two devices approached the original performance specification. More software was written which proved that these devices were correct except for a s-a-0 fault on pin 32, driven by an output pad, PDF in the ULA9C. However, the student design had passed all Qudos' simulation and error check procedures and, from a graphical plot (Fig. 4), the integrity of all cells in the vicinity of this pad were verified. The remaining chips were so erratic that it was felt further enquiry was precepted.

An optical microscope was used to examine one 'nearly working' and one other IC. Viewed at $40 \times$ and $100 \times$ magnification the results were quite

FIG. 5   Photograph showing processing blemishes of faulty IC.

alarming. A multitude of processing problems were manifest on the latter chip, ranging from under-etch, over-etch and chemical contamination (Fig. 5). The flaws were not confined to one area of the die although the cells reserved by Qudos for the installation of their test circuit (added at manufacture) did appear relatively free from defects. Obviously this explained the aberrant behaviour of three of the batch but the s-a-0 fault could not be evinced from a visual examination of the chip. As expected, processing blemishes, where observed, were centred on unused cells and all the metal tracks associated with PDF were visually traced back through the die correctly and the physical layout corresponded exactly with a previously generated plot.

In common with many low volume custom semiconductor vendors, the only criterion Qudos use to determine yield, on economic grounds, is whether the 'drop-in' test circuit functions properly. On this basis, a die is bonded, packaged and supplied as a 'notionally correct' IC. A further four chips, again at no charge, were sent by Qudos with similar results; two displayed random faults, two had the same problem with pad PDF. Another six ICs were processed by Qudos who had a mutual interest in locating the source of the problem. Again the s-a-0 error was exhibited by just two devices.

Some further investigation, in consultation with Qudos, finally clarified the situation. At normal magnification PDF appeared identical to all other pad drivers configured for tri-state output. However, using 'LED' at maximum scale, two short lengths of metal were apparent. Removing the 'TRI PAD' peripheral overlay showed that the students had inadvertently left some tracks during manual layout which were obscured by the metal bonds of the tri-state pad circuit. Even though no metal was actually shorting, this contravened the ULA9C design rules and the 4 µm gap between the base and emitter was bridged during manufacture as can be seen from Fig. 6. From the Ferranti design manual[3] this transistor formed the data input to the tri-state pad which was shorted directly to 0V via the grounded emitter.

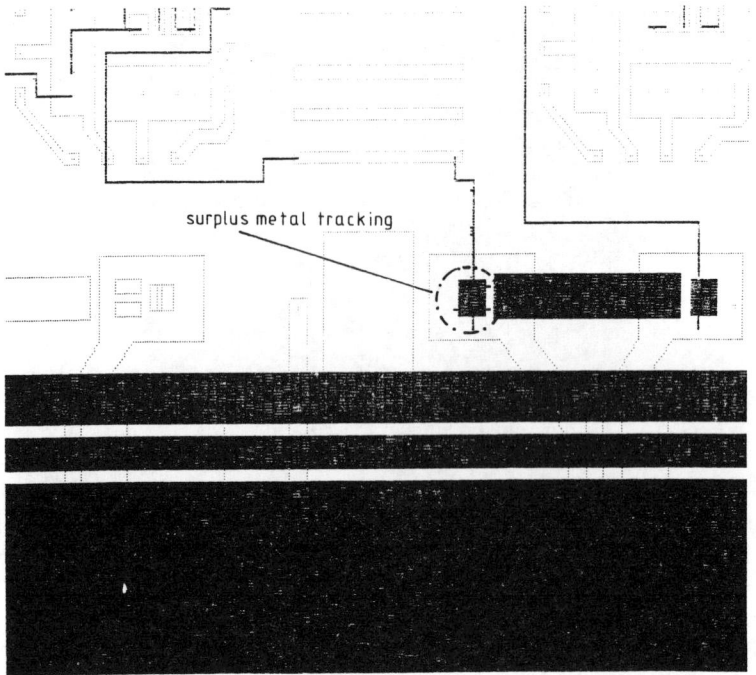

FIG. 6   Magnified plot of PDF with tri-state driver.

6   CONCLUSIONS

As a vehicle to expose students to ASIC design techniques the group project scheme proved extremely successful. Apart from technical considerations, the participants gained an insight into managing a semi-custom based project. For introductory work, the Qudos MINICHIP educational chip design package is excellent. Being less powerful than professional software, students find it easy to learn, apart from some inevitable confusion between circuit nodes and the BASIC program variables that access them in the HDL.

However, unless trivial designs are fabricated, typically to MSI scale such as shift registers or counters, MINICHIP becomes untenable for real custom chip applications. Project specifications must be carefully planned at the outset otherwise students will quickly run into difficulties. A top end figure of say 150 NOR gates should be taken as a practical design target for student projects although Qudos do offer to support multi-project submissions for a single '9C' implementation. Memory constraints of the machine and the limited nature of the software can cause problems, particularly during layout where no hierarchical compression is possible.

It was evident that the project generated tremendous enthusiasm amongst the students. A side-effect was the feedback given to other subjects such as Electronics and Management, for example. The presence of gate propogation delays in digital logic is emphasised when observed as an 8ns glitch during simulation. Final yield of the finished chips proved poor. Less than 20% of the available cells were used in the ULA9C and of the fifteen chips evaluated, nine were sub-standard due to fabrication problems. No devices conformed to specification. An internal short-circuit was discovered within the vicinity of a pad driver cell transistor, that caused an output pin to be permanently low.

ACKNOWLEDGEMENTS
The author wishes to acknowledge the contributions made by Dr. N. J. Poole and Mr. A. Pennington to the group project effort.

REFERENCES
[1]   Wallace, B., 'Introduction to image processing algorithms', *BYTE*, pp. 169–186 (March, 1987)
[2]   Qudos Ltd., *Qudos Logic Array Design Software Manual QED2* (1986)
[3]   Ferranti Semiconductors, *ULA Design Manual A/F002* (Nov., 1981)

ABSTRACTS–ENGLISH, FRENCH, GERMAN, SPANISH

ASIC technology appreciation: a 'hands-on' approach
With the aim being to expose large numbers of degree students to the practical details of a semi-custom circuit design cycle, a case study of a second year degree group project, which resulted in custom chips being manufactured, using the QUDOS 'MINICHIP' gate array design suite is presented. Administrative objectives are discussed together with an analysis of the chip samples received.

Appréciation de technologie ASIC: une approche pratique
Dans le but d'exposer à un grand nombre d'étudiants les détails pratiques d'un cycle de conception de circuits pré-traités, cet article présente un cas type de projet d'un groupe d'étudiants de deuxième année; ce projet a été conçu en utilisant l'ensemble de conception de réseaux prédiffusés QUDOS 'Minichip' et a abouti à la fabrication de puces dédicacées. Les objectifs administratifs sont discutés de même que l'analyse des échantillons des puces.

ASIC-Technologie-Wissen: ein Hands-On-Zugang
Mit dem Ziel, großen Zahlen von graduierten Studenten die praktischen Details eines Semikunden-Schaltkreisentwurfszyklus zu vermitteln, wird eine Fallstudie eines Gruppenprojektes für das Examen im zweiten Jahr präsentiert, das zu gefertigten Kundenchips führt und bei denen das QUDOS Minichip-Entwurfssystem verwendet wird. Administrative Gesichtspunkte werden zusammen mit einer Analyse der erhaltenen Chipproben diskutiert.

Apreciación de tecnología ASIC: una aproximación 'Hands-on'
Con la intención de exponer a un gran numero de estudiantes de grado los detalles prácticos del ciclo de un diseño de circuito semi-Custom, estudio de un caso del proyecto de un grupo de segundo año de grado, cuyo resultado fue un 'custom chip' fabricado, se presenta un diseño adaptado utilizando el QUDOS 'minichip' gate array. Se discuten objetivos administrativos junto con un análisis de las muestras del chip recibido.

USING SILVAR-LISCO FOR LOGIC CELL ARRAY DESIGN

MARTIN BOLTON and DAVID MILFORD
Department of Electrical and Electronic Engineering, University of Bristol, England

1 INTRODUCTION

We have been using SL-2000 in undergraduate design exercises for three academic sessions. We now believe we have evolved a method of use which fits with our overall course objectives and allows the maximum benefit to be achieved in the limited time available. This paper will present the background to our CAD course development and show how the use of programmable gate arrays has been an essential element in this.

All of our second year students undertake two group design projects. These projects extend over a period of five weeks with three three-hour sessions per week. In the running of these projects we rely on the assistance of the Department's Visiting Industrial Fellows, senior personnel who have undertaken to devote a certain amount of their time to becoming involved in departmental activities and assisting with course development. Students work on these projects in groups of 4 or 5, and therefore have to divide up the design task and coordinate their work. At the start of the task, the Visiting Industrial Fellow will assist in the presentation of the problem to the group. In some of the projects, interim reviews are held, where students present their outline designs and problems can be discussed. At the end of the design project, all students are required to participate in a final presentation and demonstration. Again the Visiting Fellow will assist with assessment, participate in the post mortem debate, and usually offer a prize to the best group.

In our Computer Systems Engineering B.Eng course, one of the design projects is a digital system design using the Silvar-Lisco CAD tools. The emphasis is on the design of a complete and working digital system, rather than on becoming expert on a limited part of the CAD system. The exercise is seen as a complement to our digital systems teaching, which emphasises planning, top-town design and design for test. Also included is laboratory work with programmable logic devices. Before the design proper, students have three sessions with the CAD system to learn the basic operations so that they can start their design work more rapidly.

We have, over the three years, learnt what type and size of problem is appropriate for this project. At first, we posed problems which were too large, which the students found interesting and tractable, but too time-consuming. As a result, there was not time to fully verify the designs. Also, some students became very involved in the task and spent too much time on the work for their own good. Examples of these designs were the serial interface of the 68HC11 microcomputer and a serial coprocessor for the TMS320. Our current view is that a design of about 1000 gate complexity is the limit. This means that the ideal problem specification is not very complex; this gives some students (who revel in complexity!) the impression that the task is trivial. However, they all admit, at the end, a tendency to underestimate the difficulty.

Until this year we have aimed for fabrication in gate arrays by the MCE Falcon service, and have had correctly working chips returned. While accepting that a physical implementation is an essential component of such exercises, we feel there are problems in using gate arrays in undergraduate design projects.

2   DRAWBACKS OF GATE ARRAYS

A student design chosen for fabrication needs verification and any 'loose ends' tying up. This takes up support staff time and adds delay. If all goes well, the fabricated chip will arrive before the end of the academic year, but at a time when the students have moved on to other things and are probably also thinking about their exams. It is thus difficult to recreate the initial enthusiasm. Added to this is the fact that many students doing the exercise come to the conclusion that their design will never be fabricated — we obviously can't afford to make them all. These factors work against our efforts to encourage the students to consider the constraints of chip resources and testing in their designs. A typical response is: 'We know that our design is unlikely to be made, so why should we not put in the extra functions which cause the gate count to exceed your arbitrary limit?' Also, there is less incentive to simulate as thoroughly as necessary. A further problem arises if layout and routing are not performed locally: only approximate performance figures are available without back-annotated delays.

When the Logic Cell Array family of programmable gate array chips was introduced by Xilinx in late 1985, it appeared ideal for student design projects such as the one we run. It is a chip which has a gate array-like structure but with RAM-programmable cell functions and interconnections. With this technology, a design can be implemented and tested within the timescale of the design project and reworked if necessary. The problem was that the development system supplied, XACT, is basically a physical design system, and not convenient for higher level logic or block diagram design. However, it was not hard to see how this system could be linked to a logic-design system such as Silvar-Lisco. We therefore undertook this task, with the support of Monolithic Memories (now a subsidiary of Advanced Micro Devices), who also manufacture and support the Logic Cell Array devices. Before discussing our experience with this device, we will briefly outline the features of the Logic Cell Array (LCA) and of the software we have produced.

3   THE LCA

The LCA is a CMOS field-programmable logic device which offers considerable functional complexity and flexibility. Fig. 1 shows the organisation of the

94

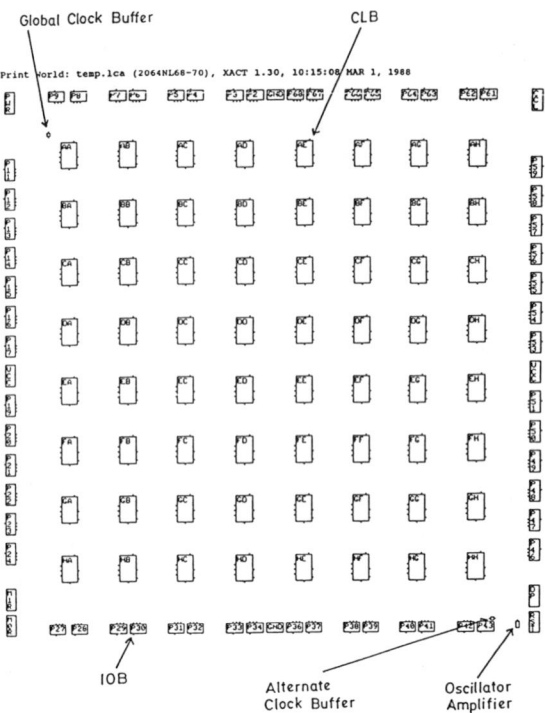

Global Clock Buffer   CLB

IOB

Alternate Clock Buffer   Oscillator Amplifier

FIG. 1   Organisation of the LCA.

device which is composed of three basic user-programmable elements. Each of the Configurable Logic Blocks (CLB) in the array can be programmed by the designer to provide a specified logic function. Around the edge of the array are a number of I/O Blocks (IOB) which provide an interface to the external device pins. The CLB and IOB architectures are shown in Figs. 2(a) and (b). Signals are routed through the array by a network of programmable interconnections. The configuration of all these programmable elements is established by writing a bit pattern to static memory in the device; the LCA has a number of dedicated pins which permit various configuration modes including automatic loading from ROM at power-up. There are minor differences between the 2000 and 3000 series LCA architectures but otherwise the range of devices varies only in terms of speed, size of the CLB array and packaging. The manufacturers' literature[1,2] should be consulted for further details.

A device of this complexity clearly cannot be programmed without computer assistance. This function is fulfilled primarily by the XACT development system which runs on the IBM PC and compatible machines. At the centre of the system is a menu-driven graphics editor which presents the designer with a low-level view of the LCA architecture; layout and configuration of CLB and IOB elements is performed using a mouse in conjunction with a small amount of keyboard input. There is also a library of macros allowing preconfigured

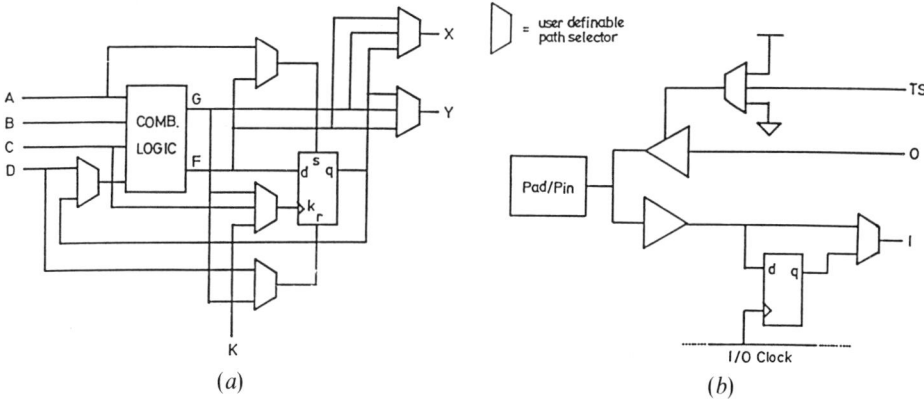

FIG. 2   (a) CLB architecture,   (b) IOB architecture.

groups of CLBs to be fetched and placed within the array. A variety of supporting routines includes provision of automatic routing, calculation of routing
delays and generation of the configuration bit stream for downloading or
EPROM programming. Configuration of the device using XACT thus enables
the designer to obtain maximum utilisation of the LCA resources but is time
consuming and somewhat error-prone.

To overcome the problems which accompany design at a physical level,
Xilinx and AMD/MMI have devised an interface which enables designers to
work at a higher level using schematic entry. The interface is based on the LCA
External Netlist File (XNF) which is an intermediate circuit description using
low-level logic primitives. For each schematic entry system there should be
library support for the LCA and a translator to generate an XNF description.
The program XNF2LCA is used in the XACT system to create an LCA file
which is then submitted to an automatic place and route program (APR) to
optimise the layout. Finally, the XNF file can be annotated with net delays
providing sufficient data for a complete simulation of the design.

4   THE SILVAR-LISCO INTERFACE
The SL-2000 suite of programs, supplied under the ECAD initiative, is used
extensively in our undergraduate courses for electronic design. To support
LCA design we have constructed a symbol library for the 2000 series LCA
family based on the macro library defined by AMD/MMI. The Silvar-Lisco
interface to XACT is summarised in Fig. 3.

Using CASS for schematic capture, designs may be entered in the normal
way making natural use of structural hierarchy. Apart from simple gates and
bistables the designer has access to a range of higher level components including multiplexers, decoders and counters. The design procedure is therefore
similar to that used for any semicustom implementation. There are additional
features which are peculiar to the LCA implementation; the library includes an
explicit CLB symbol which may be configured using the component attribute

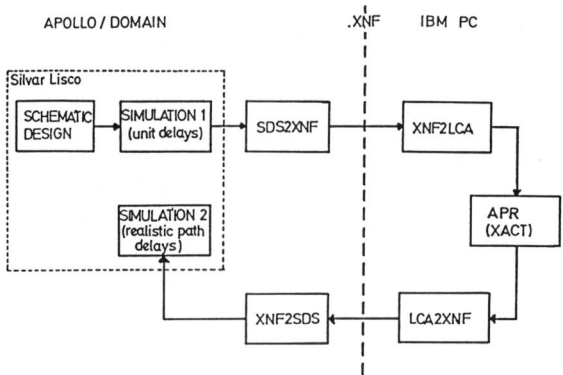

FIG. 3   Silvar-Lisco interface to X ACT.

facility and a system of layout constraints is provided as an option to assist partitioning of the target LCA. Following CASS, the circuit netlists are extracted from the schematics (using NLE) and the design flattened (using HIDEX). During the latter process the library macros (whose structure is not visible to the user) are decomposed into low-level XNF primitives.

Generation of the XNF circuit description is now possible. The translator SDS2XNF has been written in Pascal using EARS routines (the Silvar-Lisco 'Engineers Access Routines') to extract the necessary data directly from the design database. Before creating the XNF file the translator checks that various implementation-specific design rules have been satisfied, for example that connections to IOB components are legitimate. It has been found advantageous to trap basic design errors at an early stage before they are discovered in XACT. Where explicit CLBs have been used, the translator also encodes configuration details to enable subsequent simulation.

Before transferring the XNF file to the XACT system for layout, the functional behaviour of the design may be verified using HELIX. At this stage a unit-delay simulation is used since the routing delays (as yet unknown) are particularly significant in determining the final performance of the device. We have written HELIX models to support each of the XNF primitives

The remainder of the process using XACT is very straightforward. XNF2LCA takes the list of XNF primitives and assembles them into CLB configurations taking account of any constraints which might have been specified earlier. Explicit CLBs in the original design remain intact during this process. The resultant LCA design file is then processed by APR which uses a technique of simulated annealing to optimise placement and routing. This program can take a considerable time to execute but we have experience of several successful layouts with over 90% CLB utilisation. At this stage all the routing delays have been determined and a procedure has been defined for back-annotation via the XNF file to SL-2000 for final simulation. We expect to implement this in the near future.

5 TESTING THE DESIGN

To facilitate testing of LCA designs an 'LCA Demo Board' is available from AMD/MMI. This is delivered with a 2064 LCA device (8 × 8 CLB array) in a 68 pin socket and additional circuitry which allows it to operate as a stand-alone system. The board also carries a socketed EPROM which is programmed with four demonstration LCA configurations. User-defined configurations can be downloaded from XACT or programmed into a substitute EPROM. Direct connections to the device I/O pins are brought out to a 64 pin edge connector enabling a simple hardwired interface to be made to test-equipment.

In our design exercise we use a Rohde and Schwarz Logic Analysis System (LAS) to stimulate the LCA and analyse its response. The signal generator is programmed with test vectors identical to those used during simulation under HELIX so behaviour can be readily compared. In the future we expect to operate a system whereby test vectors and simulation results are ported to the LAS, which can then be programmed to provide a completely automatic test facility.

6 EXPERIENCE

We have used a preliminary version of this system with the 1987/88 class. The design this year was devised and presented to the students by David Burrows of Plessey Research. It is a subsystem which accepts 4-bit elements of an image serially and determines whether the central element is 'significant' in comparison with the surrounding ones. This design had been previously built using some 6 PAL devices and thus was of suitable complexity for the smallest LCA. A block diagram of the spatial filter is given in Fig. 4. The blocks perform the following functions:

INBUF   input buffer

MXSBB   generates a running maximum

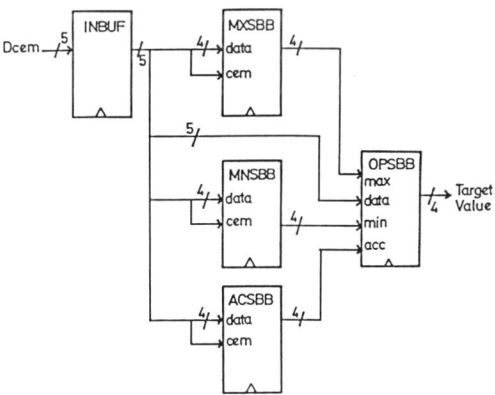

FIG. 4   Block diagram of the spatial filter.

MNSBB   generates a running minimum
ACSBB   generates a running average
OPSBB   calculates (max − min + acc) and updates output if less than central element value.

In order to avoid unforseen difficulties with the second year class, this design was then made the basis of a third year project[3]. Full details of the design may be found in the report.

The design class students were presented with the problem and given a block diagram, roughly corresponding to PAL device level. The interesting questions to be answered were: How fast will the design go? Can it be made to fit into an 8×8 LCA? What is the best way to do the arithmetic?

All groups were able to get to the layout stage, with one producing both the fastest and most economical design. Because of the 'instant fabrication' and testing possible, more than one iteration could be made. Students learnt the practical way the importance and difficulties of testing, and, in fact, test vectors from one group identified a problem on another group's 'working' design!

7 CONCLUSIONS

The important question to answer is: 'Did we meet our educational objectives?' Of course, the risk in answering this question is that we can adapt the objectives to match the reality. However, with this in mind, we will list our objectives, and attempt to judge to what extent they were met in practice.

Development of team skills   This is an objective of all of our design projects. To the degree that we achieve it in any of them, we achieve it here. Much depends, however on the personalities of the students in a group; a dominating member can sometimes be responsible for 90% of the work.

Appreciation of the importance of well-planned designs   This is perhaps the easiest one to achieve. The training of the Computer Systems Engineering students makes this almost second nature.

Design for testability   We can only achieve a limited amount here. There is not time in the second year of the course for extensive teaching of the techniques required. Nevertheless, with the limited designs undertaken, the commonsense approach is often sufficient.

Contact with semicustom technology   This is obviously achieved through use of the LCA, a programmable semicustom component. However, we try to hide the details of the particular device where possible, as this merely adds complexity without revealing any new principles.

Effect of physical constraints on performance and architecture   This is an objective which we have had difficulty in meeting hitherto, for the reasons given in Section 2. But now we feel that we can achieve something here. While the

LCA has different timing characteristics to standard gate arrays, they are qualitatively similar.

Capabilities and limitations of CAD This we are sure we achieve! We find the presence of a Visiting Industrial Fellow particularly useful here to field the complaints of the students about slow and sometimes unreliable systems. The response is that this is a taste of the real world!

In conclusion, we feel that with experience and fully debugged software we can achieve more of our aims with this type of device than with conventional gate arrays. To reiterate, these aims are the teaching of top down digital systems design for semicustom ICs.

RFERENCES

[1]   Xilinx Inc., *The Programmable Gate Array Handbook* (1986)
[2]   Advanced Micro Devices, *LCA Application Handbook* (1987)
[3]   Marshall, D. P., 'Evaluation of the logic cell array design system'. *B.Eng Project Report*, Department of Electrical and Electronic Engineering, University of Bristol (June, 1988)

ABSTRACTS–ENGLISH, FRENCH, GERMAN, SPANISH

Using Silvar-Lisco for logic cell array design
Instead of gate arrays, the logic cell array (LCA) has been used as the implementation medium in an undergraduate design exercise. Software has been written to link the SL-2000 database to the LCA layout and routing system (XACT). The result has been improved student motivation and a higher proportion of completed designs.

Utilisation de Silvar-Lisco pour la conception de réseaux à cellules programmables
Au lieu de réseaux prédiffusés, des réseaux de cellules logiques (LCA) ont été utilisés comme moyen d'implémentation dans un exercice de conception à l'usage d'étudiants ingénieurs. Un logiciel a été écrit pour relier la base de données SL-2000 à la disposition des LCA et à leur routage (XACT). Le résultat a montré une plus grande motivation des étudiants et une plus grande proportion de projets complétés.

Nutzung von Silvar-Lisco-Entwurfssoftware für programmierte Gate-Arrays
Anstelle von Gate-Arrays wurde das Logic-Zell-Array (LCA) als Implementationsmedium bei einer Entwurfsübung für undergraduate-Studenten genutzt. Software wurde geschrieben, um die SL-2000 Datenbasis mit dem Layout- und Routingsystem (XACT) zu verbinden. Die Ergebnisse haben die Motivation der Studenten verbessert und den Anteil kompletter Entwürfe erhöht.

Diseño de Gate Arrays programables utilizando Silvar-Lisco
En lugar de gate arrays, se ha utilizado logic cell array (LCA) como implementación media en un ejercicio de diseño para no graduados. El software se ha escrito para unir el SL-2000 base de datos con el LCA fichero de salida y el sistema trazador (XACT). El resultado ha sido una mejora en la motivación del estudiante y una mayor proporción de diseños finalizados.

THE ROLE OF THE FULL-CUSTOM DESIGN EXERCISE IN TEACHING

RICHARD G. FORBES and BARRY M. COOK
Department of Electronic and Electrical Engineering, University of Surrey, England

The authors have, for the past three years, been involved in practical design teaching using the ISIS full-custom CMOS design package marketed by RACAL-REDAC, installed on RACAL's proprietary platform (the V800 workstation).

The 'obvious' use for such a system is in the context of undergraduate final-year projects: given a suitable degree-course structure, a student can design, simulate and layout a digital integrated circuit of the complexity (say) of a traffic-light controller, and have the chip fabricated and returned for testing within the academic year. A description of this type of design activity will be found elsewhere in this issue.

There are, though, time constraints on the use of specialised equipment. So a limitation of this kind of usage is that only the relatively small number of students who wish to do (or can be accepted for) IC design projects gain experience of full-custom techniques. However, we believe that all modern-day engineers should have some appreciation of such techniques, and, con-sequently, we think it is best to give *all* students some hands-on experience of our full custom system. Thus, we have introduced a 6-hour IC design exercise into our normal 2nd-year undergraduate laboratory programme. In general terms, we hope that the exercise will contribute to understanding the IC design process, will 'demystify' those 7400 series little black boxes, and will contribute to the making of a link in students' minds between semiconductor physics and electronic circuit behaviour. This note summarises our thinking and experi-ences with this type of exercise.

The ISIS layout methodology employs an artefact-based editor (as opposed to the more common 'polygon editor'), and has interactive design-rule-checking and a number of other features which — in our view — make it particularly suitable for undergraduate teaching. It also uses a Hardware Description Language (HDL) as its normal mode of circuit design capture.

Students are required to do a couple of hours preparation in their own time before doing the design exercise. They are provided with notes on (and examples of) the use of ISIS HDL and the ISIS Simulator, and as part of their preparation they write HDL descriptions of a small hierarchically organised digital circuit, such as a half-adder. Then, during the exercise proper they enter these descriptions into the design system, simulate them, and then layout and validate their design.

The design capture and simulation phases of the exercise are carried out at an auxiliary terminal, and the layout and validation phases at the workstation. The workstation session involves: some demonstrations; an hour learning the CMOS layout editor by going through a worksheet; an hour to lay out a NAND-gate and perform first-level design validity checks; and an hour to use the place and route facilities to build up the half-adder from this NAND-gate and a pre-provided inverter, and to do design validity checks.

Thus, in the course of the exercise the student has gone through the whole IC design process, albeit for a very simple circuit, and has a design that would in principle fabricate correctly. Obviously, the electronic design component is trivial — though the discovery with the simulator that even simple circuits can show glitches comes as a surprise to some students. But this choice of a small hierarchical design is deliberate: students already understand the digital logic involved; the design can be done quickly; and such a design illustrates the hierarchical features of the system.

In a real design, of course, one might choose to construct the half-adder directly out of transistors in a circuit not decomposable into gates. At the end of the exercise we show students a half-adder built in this way, in order to illustrate the difference between a 'TTL design style' and a 'CMOS design style', and to demonstrate the saving in silicon space that can be obtained.

Students normally work in pairs. As one might expect, we find that pairs usually are better than either one student working alone or three students working together. With two workstations and two auxiliary terminals we can accommodate a group of 8 students. So with two lab sessions per week and a nine-week circus, we can accommodate our year-group of 120 students.

We also find that effective teaching requires the demonstrator to keep a closer watch than is normal in laboratory classes over what the students are doing, and in particular how they are using the mouse and interacting with the graphics screen. Eight inexperienced students are sufficient to keep one demonstrator quite busy.

Student reaction to the exercise is generally good, though a small minority find it lacking in intellectual stimulation. We find that the need to describe circuits in a hardware description language causes no difficulty for electronic engineering students.

Obviously the exercise reinforces lecture materials on the IC design process. But we would argue that it also contributes to the required 'EA1/EA2'* component of engineering formation, by giving students experience of a modern engineering CAD tool.

---

* *Engineering Applications*, as recommended by the Committee of Inquiry into the Engineering Profession, London, chaired by Finniston, H.M.S.O. Cmnd 7794 (Jan., 1980).

102

ABSTRACTS–ENGLISH, FRENCH, GERMAN, SPANISH

The role of the full-custom design exercise in teaching

It is argued that one useful role for a full-custom design system is to provide all electronic engineering students with a small taste of the use of CAD in the full-custom design process, as part of the normal laboratory programme of practical activity.

Le rôle de la conception de circuits dédicacés dans l'enseignement

L'argument présenté ici est de montrer qu'un rôle utile de la conception de circuits dédicacés est de donner à tous les étudiants en ingéniérie électronique une première approche de l'utilisation de la CAO dans le processus de conception de ces circuits, en tant que programme normal d'activités de laboratoire.

Die Rolle der Entwurfsübung von nichtverdrahteten integrierten Schaltungen im Unterricht

Es wird gezeigt, dass eine nützliche Rolle eines Entwurfssystems für nichtverdrahtete integrierte Schaltungen darin besteht, dass alle Studenten der Elektroniktechnik eine kleine Kostprobe der Benutzung von CAD in diesem Entwurfsverfahren erhalten, als Teil des normalen Laborprogramms praktischer Arbeiten.

El papel del diseño full-custom en la enseñanza

En este articulo se presentan argumentos sobre la utilidad del papel que juega un sistema de diseño full-custom, para proporcionar a todos los estudiantes de ingeniería electrónica un primer conocimiento sobre el uso del CAD en los procesos de diseño full-custom, como parte de un programa normal de laboratorio en el que se realiza una actividad práctica.

FULL CUSTOM VLSI DESIGN FOR COMPUTER SCIENTISTS

ADRIAN JOHNSTONE
Department of Computer Science, Royal Holloway and Bedford New College,
University of London, England

INTRODUCTION
This paper describes the computer science oriented course on VLSI design at
Royal Holloway and Bedford New College (RHBNC). The primary aims are to
enable students with very limited electronics knowledge to design and fabricate
full custom VLSI devices, and to introduce them to the use of high level design
tools. During the academic year 1987–88 twenty-two students sat the course.
Seven of these produced systems-level projects using the ELLA modelling
language with small amounts of full custom layout implemented. These
included image processing hardware and a microcoded machine emulating a
subsystem of the MC68000 instruction set. The others produced small
subsystem-level projects with fully-laid-out and design-rule-checked geometry,
of which three were fabricated using ES2's 2 micron e-beam processing.

The success of the course is totally dependent on the power of the design
tools used and (for fabrication) the support from the University of London's
Microelectronics Design Centre based at University College.

I shall describe the resources and tools used, the format of the course
(especially in view of the stringent timetable imposed on the fabrication-
orientated projects), three projects in the form of case studies and developments anticipated for the year 1988–89

HISTORICAL BACKGROUND
Practical undergraduate VLSI teaching at the University of London began
with a pilot scheme at Queen Mary College in 1984–85. During that year Royal
Holloway College (now renamed Royal Holloway and Bedford New College
after the University of London restructuring) invested in the ISIS design system
developed by Inmos and marketed by Racal Redac as part of the Visula
package. The software ran on a dedicated MC68010 based Front End Processor (FEP) which required a VMS based VAX as a file server. A course was
run that year using a VAX 11/780 in the Computer Centre as the host. The high
demand for layout time led to round-the-clock working as the fabrication
deadline drew near.

Subsequent support for ISIS from the ECAD initiative led to significantly
reduced software maintenance costs and this, along with an injection of capital
by the Court of the University allowed us to purchase two VAXstation II GPX

colour workstations. These operate in a cluster with the original microVAX/FEP and a further departmental machine which is available for batch simulations. Hence three layout screens and about three and a half microVAX processors are dedicated to the course.

STUDENT PROFILE

The course is a third year option in the Computer Science Department. Electronic Physics and Electronics students are also allowed to sit the course, subject to loading restrictions on the equipment. In 1987–88 about a quarter of the students were non-computer scientists.

In extreme cases our students may have only O-level physics, and electronics experience limited to the construction of simple TTL designs on patchboards. Most of the students have no experience of analogue or large scale digital design. However, they are all competent programmers and in the case of the computer scientists, have experience with functional languages.

AIMS

VLSI specialists cover the range from process engineers, through layout experts to systems designers and semi-custom users. In a single course it is neither possible nor desirable to attempt to cover too much ground. This course aims to produce computer specialists with an in-depth knowledge of the implications of full custom VLSI design so that they can judge whether an algorithm is readily implementable in chip form. In particular, the well known industry adage that in VLSI design 'wire is expensive and the transistors come for free' is used to reinforce the contrast between software systems (in which computation is expensive and data storage cheap) and integrated circuit hardware in which large scale data storage can dominate costs over computation.

The students learn to use behavioural, switch level and analogue simulators, and perform layout exercises, in some cases on a scale suitable for fabrication. The software tools used are very hierarchical and promote a design methodology that reinforces the software engineering practices taught in other computer science courses.

COURSE FORMAT

The course runs for one year comprising 25% of the third year timetable but fully laid out designs must be produced by the end of the first term, i.e. within about eleven weeks of the start of the course. This allows fabricated designs to be returned to the students in time for test results to be included in project reports. Projects not intended for fabrication run throughout the year. Overall, half of the course time is allocated for project work and half for formal lectures. An average of one and half hours of lectures are given per week.

Assessment is by a single paper examination and by project work. Equal weight is given to projects and examination performance. Although the course is one of the more intensive third year options it is clear that the students are highly motivated, and their performance is often better than in other courses.

TEXTBOOKS

Until recently, there has been a dearth of textbooks available that emphasize CMOS technology. This course concentrates solely on CMOS because electronically naive students can easily design static CMOS gates using a simple switch level model of the transistor. The primary textbook is Weste and Eshraghian[2] with Mukherjee[2] as a backup. Glasser and Dobberpuhl[3] is recommended for students with an interest in the physical aspects of device design and Maly[4] is used to illustrate device processing. We are considering using the new text by Pucknell and Eshraghian[5] for future courses.

TIMETABLE

Students are introduced to static logic implementation via a switch level model of the transistor. Project topics are chosen in the second week after preliminary simulation exercises. There follows a crash course in basic electronics which allows the students to predict rise and fall times and the drive capabilities of their circuits. The ISIS simulator is used as an experimental tool in this part of the course. Practical sessions allow the students to model the behaviour of circuits which have been treated theoretically in the lecture room.

In the following weeks of the first semester lectures on flip-flops, arithmetic circuits and register design are interspersed with practical sessions explaining the use of the design tools. Students aiming for fabrication progress rapidly on their projects up to the end of the first semester.

The needs of students aiming at fabrication conflict with those of the students on systems-level projects. The first group need to ascend a very steep learning curve quickly to meet their deadline. As well as learning enough electronics to make their designs work, they must get to grips with the layout tools. The second group wish to explore the wider aspects of VLSI design and concentrate on the ELLA language. Since the lecture timetable is limited, most of the first term has to concentrate on low level details, to the possible detriment of the systems level project students. The use of high level layout design tools mitigates this problem somewhat.

In the second semester more advanced topics are treated, including various forms of dynamic logic, use of PLAs and design for test. During the first semester, certain design parameters such as width of power supply rails and substrate tap spacing are described using rules of thumb taken on trust by the students. The underlying physical phenomena are explained qualitatively so as to justify the approximations made earlier. However detailed electrical analysis is not attempted, and students are not required to remember or derive formulae in the examination.

TOOLS

ISIS

ISIS comprises a BCPL-like hardware description language (HDL), a hybrid simulator (Hy LAS) capable of modelling different parts of a design at logic and circuit levels within a single simulator run, and a graphics package with a sticks based symbolic CMOS editor, a polygon editor and a schematics editor. The system is very hierarchical and aims to decouple the logic design stage from the layout design.

Hardware description language

Designs are entered using ISIS HDL. Since our students are proficient programmers this is a natural way to work. ISIS does, in fact, provide facilities for schematic capture, and subsequent automatic generation of HDL, but considerable annotation has to be applied to schematic diagrams and they, of course, require considerably more effort to modify than a compact textual description of the circuit. In addition, use of schematic entry requires time at the graphics screen which may be in short supply due to the demands of students laying out their designs.

Blocks of hardware in HDL are represented using a MODULE which roughly corresponds to a procedure in a programming language. Some modules are primitives supplied by the system such as transistors and resistors. Primitive modules are not decomposable in much the same way that FORTRAN intrinsics or Pascal predefined routines are monolithic. Other modules are formed by connecting modules together in a hierarchical fashion.

Each HDL module has a parameter list which describes its connections. HDL modules map onto rectangular, non-overlapping areas of silicon at layout time. All the components (and their interconnects) listed in the HDL module must be contained within the layout module. Connections to the module correspond to wires crossing the boundaries of a layout module.

The HyLAS simulator

Simulation in ISIS proceeds directly from the HDL so logic and circuit level design can proceed almost independently of the geometry design. This allows the teaching of layout level details to be postponed, and as a result students are able to make significant progress on their projects early in the course.

The mixed-mode capability allows the speed of switch level simulation to be traded off against the accuracy of analogue simulation. In practice the analogue simulator was preferred. This was because it was inherently more accurate, and because some of the early example circuits such as the Suzuki adder[6] are not handled well by switch level simulators, which led to a loss of confidence on the part of the students. The time penalties were relatively slight because the circuits designed by the students were usually limited to a few hundred transistors, and considerable computer power was available to them for simulation.

The layout editor

The CMOS symbolic editor features on line design rule checking so that students are immediately informed of errors. They do not have to wait for a batch design rule check to find layout errors. In a very real sense, the software makes up for the student's relative ignorance of the underlying electronic and physical properties of the devices. The layout editor also performs network connectivity checks against the circuit specified in the HDL, which acts as the master specification of the design. There is no need for circuit extraction or simulation from the layout. One of the potential problems with the use of the ISIS system is that all layout checking (including global connectivity checking) is done from within the graphics package. There is no way to set up a large check as a batch job, and to screen utilisation is reduced.

ELLA

ELLA is a powerful hardware modelling language capable of describing designs at an architectural level. Nodes may be functions or parameterised macros. The key feature is the ELLA *type* which allows the range of values passed along wires to be specified symbolically. In its basic form it is analogous with the Pascal enumerated type. ELLA also provides ELLA-integers which comprise a tag and an integer value. The strong typing mechanism of ELLA ensures that a signal representing an address bus cannot be inadvertently connected to a node requiring a data input. This is in contrast to ISIS HDL which does not distinguish between signals other than to check the width of buses.

The ELLA type was found to be especially useful in modelling microcoded machines where the different fields of the microcode could be explicitly represented as interconnections between functional modules. This leads to a compact and easy to read notation.

It is often necessary to override the strong typing mechanism of ELLA, especially in structural simulations where ELLA integers may need to be converted to arrays of Booleans. ELLA provides an elegant mechanism known as the associated type which allows signals of one type to be associated with the new basic values of another type. Unfortunately, this was treated as an advanced topic treated near the end of our course. Several students constructed recursive functions that evaluated integers bit by bit to convert them to booleans. These macros were extremely expensive in machine resources, and in future associated types will be introduced much earlier in the teaching programme.

FABRICATION

The London Colleges have cooperated in a joint fabrication exercise for the last three years. The nine active departments typically produce around twenty-five designs for each fabrication run, and this enables the consortium to negotiate good prices with silicon suppliers. MCE were used as brokers for the first run and some ISIS designs were fabricated. For the last two years ES2 have produced the University's designs. Prior to the academic year 1987–88 no academic ISIS designs had been fabricated by ES2, but a commercial customer of Racal Redac provided us with a great deal of support at no cost, and the central facilities at University College were used to provide a backup check.

PROJECT CASE STUDIES

A list of some thirty projects was available ranging from a combinatorial adder to a microcoded processor emulating a subset of the 68000 instruction set. Three projects will be described here: (1) an edge detector for image processing which was modelled in ELLA and run on real image data, (2) a microcoded engine described in ELLA which has 'run' small 68000 programs, and (3) a serial multiplier which was fabricated and tested within the one year course timetable.

Case study 1: an ELLA edge detector

This project was to design an image processing edge detector and implement it in structural ELLA, i.e. code that could be translated by hand or using the Redac supplied translation program ELLA2HDL into ISIS HDL.

Images are represented as square arrays of 256 × 256 eight-bit integers. The value of each integer represents the brightness in the image at the point corresponding to the coordinates of the array element. A value of 0 corresponds to black and 255 to white. Edges in the image correspond to areas of rapidly changing intensity, or large intensity gradients. Various digital approximations to the derivative of the intensity have been proposed as edge enhancers or detectors. One of the simplest to implement (although by no means the most satisfactory theoretically) is the Roberts cross operator[7]. This scans a two by two pixel window across the image finding the difference of the diagonal pixels. The edge magnitude is then taken as the root of the sum of the square. The direction of the edge may be calculated from the arctan of the two components (Fig. 1).

One of the main aims of the student's design was to minimise pixel I/O which tends to dominate simple image processing operators. The time required to fetch all four pixels in the window from a frame store at each point in the image would be much greater than the propogation time of the arithmetic circuitry required. The design buffers an entire line of the image in a 256 × 8 bit shift register, and as a result only one pixel read is required per operator window calculation, assuming that data is allowed to wrap-round at the edges of the picture. Two eight-bit subtracters are used to form the partial sums. The square

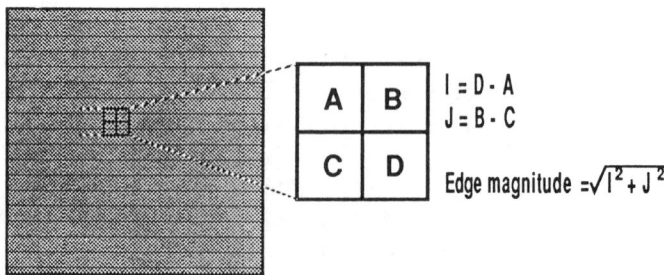

$$I = D - A$$
$$J = B - C$$
$$\text{Edge magnitude} = \sqrt{I^2 + J^2}$$

FIG. 1   The Roberts cross edge operator.

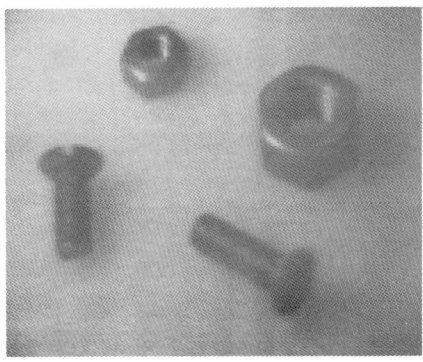

FIG. 2   Input image.

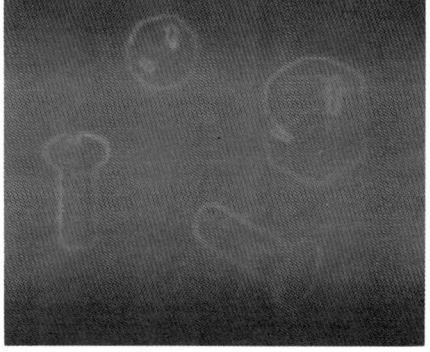

FIG. 3   Results of student simulation.

root of sums calculation $\sqrt{i+j}$ is performed using a look-up table addressed by the partial sums. Because i and j always yield a positive result, only positive values need be set up in the look-up table, however the outputs of the sub-tractor will be required in absolute form, i.e. they should be negated if less than zero. This reduces the size of the look-up table from 512K × 8 bits to 64K by 8 bits. In practice the size of the look-up table can be further reduced since many applications do not require high accuracy.

The student had to depart from the project specification in that his attempt to generate a look-up table of the required size using the ELLA RAM primitive overflowed system memory limits. As a result he wrote a recursive approxima-tion to the square root routine which was used instead of the look-up table.

As a full test of the design, a standard benchmark image from the author's research laboratory (Fig. 2) was converted into an ELLA simulation file and run through the simulated circuit. The results were converted back into image data for checking (Fig. 3). The results correspond to that calculated by a Pascal program. However, some twelve CPU hours were required to generate the image shown here. Image processing projects like this are especially satisfying for the student because the results are directly visible and are more meaningful than a long printout of simulated waveforms. The early success of this project spurred on other students in the group.

Case study 2: an ELLA 68000 subset processor
A second year assembly language and interfacing course introduced the students to microcoding via a paper processor with a load/store architecture. This project required the student to generalise the design to a processor with memory to memory capability, and implement it in ELLA. The student then had to write microcode for his processor that would allow it to emulate the 68000 instruction set. No reference was made to the internal design of a real 68000.

The first decision was to generate a subset of the 68000 instruction set capable of demonstrating real programs but without requiring large amounts of microcode to be written. The following classes of instructions were omitted: BCD instructions, 'quick' instructions such as MOVEQ and ADDQ, LINK and UNLNK, MUL and DIV and bit manipulation instructions apart from those acting on the condition codes register. In addition, only 16 bit data were supported and the MOVE instruction was limited to its most basic form. This left the basic arithmetic, logical and shift operations, compare, branch and jump, LOAD EFFECTIVE ADDRESS, MOVE and CLEAR instructions. The addressing modes of the 68000 were fully implemented along with general two effective address instructions such as MOVE.

The hardware model designed to support these instructions included all the programmer visible registers of the 68000 (i.e. eight data registers, eight address registers, a program counter and a condition codes register) along with an eight function ALU, a shifter, eight scratchpad registers for microcode temporaries, a set of instruction registers and a microcode section acting as controller. The microcode word size was 52 bits.

Since the model was meant to be structural rather than just behavioural, the design had to be specified down to the level of basic gates operating on boolean values. The inbuilt logical operators in ELLA are only defined on ELLA integers so considerable work had to be done by the student to build up his working simulation. Within the processor, latches were used for storage. The main memory was represented using the ELLA RAM primitive.

The completed 500 line ELLA implementation represents a large scale use of ELLA for structural modelling, and in principle the design could be 'pushed down' into ISIS HDL using the translator program. At this point however, the project was only half completed because the student had to write microcode to implement the chosen parts of the 68000 instruction set and test it. This proved to be very time-consuming, especially in view of the fact that the student was attempting to hand-assemble 52-bit microcode words and then type them into the simulator as a string of binary digits. In future years, this project will be supported by a microcode assembler written for another project, that will allow symbolic assembly code to be written. In spite of the difficulty of the task, the student demonstrated simple programs running on his ELLA processor.

Case study 3: fabricating a serial multiplier
In contrast to the previous two projects, this project required little design work at the systems level. The circuit for the serial multiplier was taken from Weste and Eshraghian and a pencil and paper 'walk through' done to ensure that the student understood its operation. A behavioural simulation in ELLA was then performed to check that the relationship of the clock timing signals was correct.

The ISIS HDL description of a four by four multiplier (yielding an eight-bit result) is only about 200 lines long and contains 205 transistors. This was extensively simulated at circuit level.

Laying out the design was time-consuming. The design of the leaf cells was completed quickly but global routing was very slow. ISIS includes tools for auto-routing, but at this time documentation had not been received and the students preferred to do it manually rather than experiment. In the case of this project, the problem was made much worse when close to the completion of his layout the student realised that insufficient substrate taps had been inserted in the leaf cells. Considerable amounts of reworking were required.

The resulting circuit occupied about one third of a ten mm^2 die. Two other student projects were to be fabricated on the same chip, but in the event one of the projects was withdrawn at the last moment because of design defects. Twelve samples were returned in the last week of the Easter vacation and tested using unsophisticated equipment. The circuit operated correctly, but the timing relationships were very finely balanced and reliable operation at speed has not been demonstrated. Other designs which were more combinatorial were more successful.

FUTURE DEVELOPMENTS

The Physics Department at RHBNC has mounted a first year course on Electronics for Computer Scientists for the last two years. The syllabus was specifically designed as an introduction to the analogue behaviour of CMOS digital circuits and will in future be a prerequisite for the VLSI design course. In 1988–89 the first students from this course will be taking VLSI design, and therefore the current crash course in analogue electronics will be dropped. This will enable the course to concentrate on more computer science-orientated topics such as complexity analysis for VLSI layout, formal specification and verification of hardware, and CAD algorithms and implementation.

Further funding from the University Court will increase the number of layout screens to five, and six processors will be available for simulation. Enhancements to the ISIS software will allow increased automation of the layout stage, and this, coupled with ELLA to ISIS HDL translation tools and cell libraries supplied by UCL will allow our students to fabricate systems level designs.

CONCLUSIONS

This course demonstrates that non-electronics specialists can produce working VLSI designs within an eleven week semester, although considerable effort is required on the part of students and staff. Static CMOS is an ideal technology to teach because simple models of transistor operation are close enough to reality to enable successful designs by inexperienced students. The availability of powerful graphics computers along with the possibility of fabrication motivates the students strongly, and examination results often show weak students performing better than expected. What is yet to be demonstrated is that large systems scale projects can be successfully implemented in the required timespan. Success will depend on the availability of large, proven cell libraries.

REFERENCES

[1]   Weste, N. and Eshraghian, K., *Principles of CMOS VLSI Design. A Systems Perspective*, Addison-Wesley (1985)

[2]   Mukherjee, A., *Introduction to nMOS and CMOS VLSI Systems Design*, Prentice Hall (1986)

[3]   Glasser, L. A. and Dobberpuhl, D. W., *The Design and Analysis of VLSI Circuits*, Addison-Wesley (1985)

[4]   Maly, W., *Atlas of IC technologies: An Introduction to VLSI Processes*, Addison-Wesley (1987)

[5]   Pucknell, D. A. and Eshraghian, K., *Basic VLSI Design*, Prentice Hall (1988)

[6]   Suzuki, Z., Odagawa, K. and Abe, T., 'Clocked CMOS calculator circuitry', *IEEE Journal of Solid-State Circuits*, **SC-8**, No. 6, pp. 462–469 (Dec., 1973)

[7]   Roberts, L. G., 'Machine perception of three dimensional solids', Ch. 9, pp. 159–197, in Tippet, J. et al. (eds.), *Optical and Electro-Optical Information Processing*, MIT Press (1965)

[8]   Heinbuch, D. V., *CMOS3 Cell Library*, Addison-Wesley (1988)

ABSTRACTS–ENGLISH, FRENCH, GERMAN, SPANISH

Full custom VLSI design for computer scientists
This paper describes the computer science orientated course on VLSI design at Royal Holloway and Bedford New College, the primary aims of which are to enable students with very limited electronics knowledge to design and fabricate full custom VLSI devices, and to introduce them to the use of high level design tools.

Conception de circuits VLSI dédicacés pour des informaticiens
Cet article décrit le cours, orienté vers les sciences informatiques, relatif à la conception de circuits VLSI au Royal Holloway and Bedford New College. Les buts essentiels de ce cours sont de permettre à des étudiants ayant des connaissances très limitées en électronique de concevoir et de fabriquer des circuits VLSI dédicacés ainsi que de les conduire à l'utilisation d'outile de conception de haut niveau.

Vollkunden-VLSI-Entwurf für Computer Scientists
Der Artikel beschreibt den für Computer Science ausgelegten Kurs zum VLSI-Entwurf am Royal Holloway und Bedford New College. Dessen Ziel besteht primär darin, Studenten mit sehr begrenzten Elektronikkenntnissen zu befähigen, Vollkunden-VLSI-Schaltkreise zu entwerfen und herzustellen und sie in die Nutzung von High-Level-Entwurfswerkzeugen einzuführen.

Diseño full custom VLSI para científicos en computadoras
Esta articulo describe un curso de ciencias de computador orientado al diseño VLSI impartido en el Royal Holloway y Bedford New College, cuya finalidad era conseguir que estudiantes con pocos conocimientos de electrónica, diseñaran y fabricaran componentes full custom VLSI, e iniciarles en el uso de herramientas de diseño de alto nivel.

STUDENT FULL CUSTOM DESIGN

W. A. J. WALLER, L. T. WALCZOWSKI, K. R. DIMOND and C. DAWSON
Electronic Engineering Laboratories, University of Kent at Canterbury, England

1 INTRODUCTION

In the final year of the Computer Systems Engineering and Computer Science courses at the University of Kent, students undertake a group project. For the past two years a number of groups have undertaken projects in the area of full-custom IC design. This paper describes some of the projects undertaken by final year students.

2 ARRANGEMENTS AND TIMESCALES

Groups typically range in size between 2 and 5. Students select projects in June of the second year and have the opportunity to discuss, with their supervisor, the overall aim of the project before the end of the second year. Project work commences in October and students are allocated approximately two days per week. At the start of project work students have taken lecture courses in CMOS design techniques, dealing mainly with gate and more complex logic structures. There is a further course dealing with time and loading analysis of logic structures which students take in the first term of the third year.

A major constraint is the submission dates for designs. We have used European Silicon Structures (ES2) to fabricate the designs, through the University of London Consortium. The tape submission date is early January. This imposes very strict timescales, students having to become familiar with the tools and design and test circuits in approximately ten weeks.

During the second term, after designs have been submitted, the groups then spend their time developing test rigs and test patterns ready for the return of the fabricated designs. Project reports have to be submitted early in the third term, so time is also spent on documentation. Completed designs return in May and there is an opportunity for students to test these in the days after the final examinations.

3 DESIGN TOOLS

Silvar Lisco's Structured Design System (SDS) is used for schematic entry during the early stages of projects to develop architectures and subsequently generate circuit diagrams. Once the overall design has been hierarchically decomposed to low level cells, these are designed symbolically using the *STICKS* suite of design tools obtained from GEC Hirst Research. After fleshing and compaction according to design rules obtained from the fabrication plant, the layout in CIF is fed to an in-house, menu-driven VLSI editor called *ECIF* (Edit CIF) which is used to array cells together and to perform the routing. An important feature of ECIF is its interface to a design rule checker, which not only allows the user to perform design rule checking of the layout as a menu option but also highlights selected design rule infringements on the screen enabling the user to quickly correct errors. *SPICE* is used to simulate the circuit, obtained from the symbolic layout of the cell, to check its timing and logic function. Once the cell has been fleshed out and compacted, the circuit is extracted with *MEXTRA*[1] from its mask layout, and simulated with *SPICE* to ensure circuit parasitics are taken into account. Recently, a menu-driven interface called CW (ChipWise) has been written which serves as a uniform front-end to all the design tools.

4   PROJECT DETAILS
The following group projects have been completed since the ECAD facilities were installed at Kent:

	Project title	Transistor count
1987	Programmable correlator	3000
1987	Serial pixel kernel processor	3000
1988	A/D converter	4200
1988	Frame store	11800
1988	Parallel pixel kernel processor	4500
1988	Control processor — dual port ram	11200

In addition to these, a number of postgraduate and individual projects have also been completed.

4.1   *Programmable correlator*
The correlator was designed to compare a reference word with a continuous bit stream, giving the correlation figure, at any time, between the two signals. The programmable correlator comprises a reference register, a sampling register, a comparator and an adder tree. The reference word is held in a 15 bit, serially loaded register. The sampling register is a 15 bit shift-register, which holds the previous 15 samples of the input data stream. The comparator comprises an array of Exclusive-NOR gates which compare the reference and sampling registers on a bit for bit basis. Thus for any two patterns, the comparator output provides a 1 for each bit-wise 'agreement'. The final section comprises a tree of full adders and half adders which sum the comparator outputs (unary weighted) to generate a 5 bit binary output word. Pipelining is used to reduce the propagation delay through the system; the adders are clocked, which produces a latency of 6 clock cycles at the output, but allows a clock rate of at least 5MHz, (which is the maximum frequency of the test system!).

Each comparator cell has an enable input which, when inactive, will force the comparator output low. These enable inputs are grouped so that 3, 7 or 15 adjacent comparators can be enabled at once. This configuration means that the chip can be used as a 15 bit, 7 bit or 3 bit correlator.

In order to improve testability, spare output pads were used to monitor various nodes inside the chip so it was possible to observe signals passing between the major blocks.

4.2 Video processing system

For the last academic year it was decided to adopt projects which were related. Thus these four projects formed the basis of a video processing system, the overall structure of which is shown in Fig. 1. The following sections describe in more detail the component parts of the system.

4.2.1 Analogue to digital converter

The first device in the video processing chain is a 6-bit flash analogue to digital converter. Although the pixel processing system only requires a 5 bit input word, a 6 bit converter was produced because

(a) There was room in the chip frame for a 6 bit converter and
(b) the converter could then be used in other applications.

The chip comprises 64 comparators, each fed by the analogue input voltage, VIN and a reference voltage, VREF. The reference voltages are generated by a 64 stage, on-chip resistor chain, which is fabricated in first layer metal. The comparator outputs are fed to clocked inverters which in turn drive an array of Schmitt triggers. The outputs of the Schmitt triggers go to a set of latches, the outputs of which feed an output encoder, which converts the 64 bit latch output to the final 6 bit output.

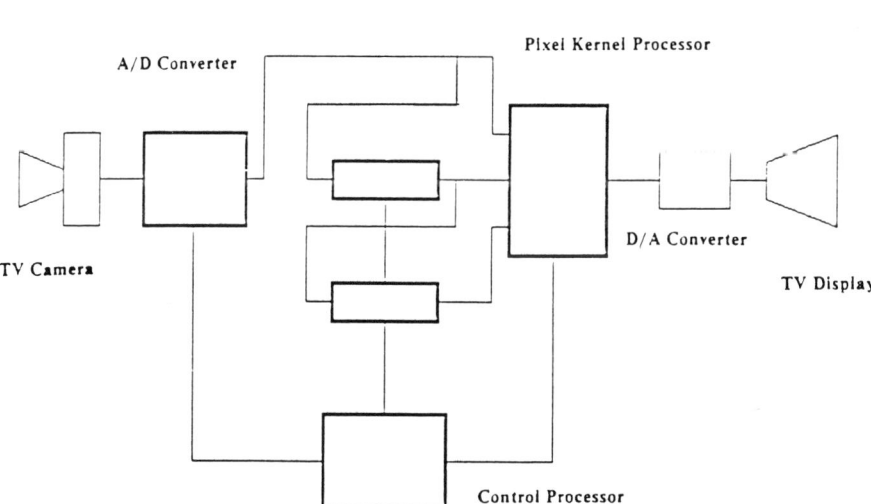

FIG. 1   Block diagram of graphics processing system (heavy lines denote modules implemented).

The comparator used is a 2 stage capacitively coupled design, after Dingwell and Zazzu[2] which is insensitive to threshold voltage variations. The coupling capacitors were formed, on chip, between polysilicon and metal.

The chip requires several clock signals, which are provided by an on-chip timing chain. The output latches can be configured to form a scan path. This allows the input of test signals in order to exercise the output encoder and also allows the comparator outputs to be scanned out. Thus an error in either of the two main sections will not prevent the testing of the other section.

4.2.2 The frame store

The frame store consists of two or more individual line store chips that were designed by two students. Essentially, it is a five bit shift register of 256 stages with a multiplexer that permits the delayed data to be tapped at intervals of 64 delays. This allows the overall delay to be tailored to the final video sampling rate that was not known initially.

A considerable part of the design time was put into optimising the shift register stage since this largely determined the storage capacity of the chip. The delay section occupied about 75% of the chip area, excluding the I/O pads. Circuit extraction was used to generate a SPICE model of the shift register and three stages were used to determine the realistic delays that could be expected of a loaded element. It was then possible to calculate the capacitive loadings presented by the clock lines and to design the clock drivers.

The *STICKS* suite did not have a floor plan editor and so size adjustment of abutting cells was usually carried out by positioning stretch lines within the cells to align the tracks. Since there are only a few different cells in this design, another approach was adopted here. The *STICKS* versions of the abutting cells were combined and then jointly fleshed out. This ensured that all tracks would align. The fleshed versions of the cells could then be separated and incorporated into the hierarchical design where they would be used repeatedly.

The design used just under 12,000 transistors and was completed by the submission deadline. Testing was performed using the output interface of a 68000 microcomputer. Test patterns in memory were rapidly transferred to the output interface lines which, in turn, were connected to the chip inputs. The resultant output waveforms were stored in a logic analyser for checking. Seven of the 12 devices were fully functional.

4.2.3 The parallel pixel kernel processor

The parallel pixel kernel processor is the main processing unit of the image enhancement system and performs a 3×1 convolution of the image data. The architecture is so arranged that three such units can be cascaded together to perform a 3×3 convolution. The five most significant bits from the flash analogue to digital converter are used to generate the data stream. The frame store holds two pixel lines of this data which together with the current scan line, provides the input to the processor. In this way, one pixel from the current scan line, plus a pixel from the previous line and a third pixel with two lines delay are presented to the pixel kernel processor at a clock rate of 5 MHz. Each set of three words represents a vertical slice of three pixels across three stored lines. Together with two other three pixel slices which have internally propagated across the other two identical processors, a 3 × 3 kernel is available for transformation according to the weighting matrix. The individual coefficients of the corresponding nine value convolution matrix are input into the chips by strobing the address lines. Depending on the values to which the convolution coefficients are set, such functions as filtering, integration and edge detection can be performed.

Each pixel kernel processor contains three multipliers and three full adders. The latter in each chip are linked together on an internal data bus and presented to the pins as a multiplexed input/output port allowing a partial result to accumulate and cascade across the three chips to give a final 13-bit sum. Each multiplier is fully parallel, the design based on the Baugh-Wooley[3] parallel array multiplication algorithm.

4.2.4 *Control processor*   The initial idea was to realise a processor which would control the operation of the above three designs. It soon became apparent that this approach was not feasible, since it was dependent upon detailed information from the other three groups. Instead, the group considered the design of a reduced instruction set computer. This would be an ambitious project, an initial design study would be undertaken and then a subset of the processor would be realised in silicon, the rest of the processor being realised using conventional components. Initial studies were made on the architecture and after a relatively short time the block diagram of the processor together with a draft instruction set was produced. Of particular interest was the way in which special instructions could be used to create the register window structure which is a feature of a number of RISC designs.

It was decided that the register file was the key component in the design since the size of this sub-block would determine the number of registers and hence the scope of the machine. The block which was designed was 128 by 8, expandable in both register length and in length of register file. The main cells which made up this design were the basic RAM cell, sense amplifiers and the row and column decoders. Considerable effort was spent in the design of the RAM cell. Initially this was designed using the *STICKS* system, but it was then realised that the simple compaction approach adopted could not yield the required density. A similar but more compact design was created using the polygon editor *ECIF*.

The remaining cells were composed using the symbolic editor, but in a few cases the automatically produced design was manually optimised by *ECIF*.

5   OBSERVATIONS

5.1   *Organisation of group projects*
At the start of the project, students work together on the algorithms and architecture. This produces a block diagram and also a provisional floor plan.

The various modules are divided between the members of the project group who then become responsible for all design stages of those modules.

The supervisor needs to be heavily involved during the architectural design phase to ensure that the design is feasible and within the capabilities of the target process. The supervisor also needs to ensure that the design is testable, so that all blocks can be exercised and observed. In this way, errors in one part of the chip will not prevent other sections from being tested.

5.2 *Benefits of VLSI group projects*

When a group of between 2 and 5 students work together, it is possible to undertake quite demanding projects. Generally, the project is partitioned and students undertake responsibility for all design stages of one or more sections of the chip. In spite of this, students become familiar with all sections of the complete system, not just their own section. They also gain valuable experience in working as part of a team on a major design. Many students find VLSI design very interesting and become strongly motivated. Using external, commercial fabrication (ES2) forces the students to work to real industrial time-scales which is good experience. An additional benefit of group VLSI projects is that the costs, which can be quite significant, are being spread across the allocation for several students.

The VLSI design activities have spawned numerous related projects. These cover hardware developments such as low cost test stations, software projects such as silicon compiler modules and have been most valuable in fostering industrial collaboration.

5.3 *Experiences at Kent University*

While students devote a lot of time to the design of the processing sections of their designs, they aften disregard the control sections. Control logic needs as much design effort as the main processing sections and supervisors need to ensure that it is not ignored. It is very easy to underestimate the amount of work required to 'finish off' a chip. The final I/O pad routing, checking and debugging phase can take several weeks.

Students, like most people, are not very good at checking their own designs. For this reason, great emphasis should be placed on the use of Electrical Rule Checkers (ERC). Also, the supervisor needs to take a keen interest in the checking. The most common errors do not occur in the circuit design or layout, but are due to errors in the interconnection between blocks. Unfortunately, whilst an ERC will detect the presence of, say, a short-circuit, it cannot highlight the inadvertently added polygon which caused it. A common problem is that students put all the various layout, schematic and simulation files into one directory on the computer and files are often inadvertently lost when someone tries to tidy up that directory.

In order to address these points, students at Kent follow the policy outlined below:

(i) Students are required to adhere to a strict timescale, which specifies the maximum time available for each phase of the design.

(ii) Students should produce good quality schematics. In the early stages this consists of block diagrams, which are supplemented, as the project develops, by circuit diagrams and floor plans.

(iii) When designing any block, all inputs and outputs should be placed around the cell edge so that when connecting blocks together, there is no need to route across cells. As soon as any block is complete and DRC clean, all its inputs and outputs are labelled. Any global connections are given local names, e.g. ALU.VSS or IPREG.CLK1. This technique minimises the risk of error during the wiring phase, but if errors do occur, the ERC report will be more helpful in tracing them. A circuit extraction program is used to generate a simulation file from the actual layout, and if the simulation results are satisfactory, the block is 'frozen' i.e. moved to a safe directory where no more edits are allowed.

This process is repeated as the blocks are built up to produce the complete design, so at any stage errors detected by ERC or DRC should be limited to the connections between blocks. The I/O pad cells are carefully labelled so that it is obvious whether a pad is an input or output pad. This reduces the risk of using an input pad where an output pad was required.

6 CONCLUSIONS

From the experience which we have gained with the projects described in this paper, we are convinced of the value of custom integrated circuit designs as undergraduate projects. Students have benefited from being able to follow the design process from inception to final testing. The discipline in having to meet a very strict deadline, together with working with meticulous attention to detail, provides a very useful experience for work in industry. Indeed, most of the students have found employment within the VLSI field.

7 ACKNOWLEDGEMENTS

We acknowledge the help of the University Grants Committee and GEC Avionics, Rochester, in helping to finance the establishment of an ECAD laboratory at the University of Kent.

We would also like to thank GEC Hirst Research, Wembley for use of their STICKS symbolic design tools, the NorthWest Laboratory for Integrated Systems at the University of Washington for their CMOS VLSI tools, and the Microelectronics Centre of North Carolina for the sources of Z-graphics, MENU and MTF without which our mask layout tools would have been considerably more difficult to write.

Finally, we must thank M. Collins, W. Corr, A. Cross, D. Fidler, A. Gollner, N. Jackson, Z. Johannes, S. Jones, P. Jordan, M. Leach, J. Lovelace, J. McClean, A. Michael, I. Morris, K. Pennington, N. Robinson, G. Smith,

P. Thompson, E. Wells, I. White and G. Wilkinson for all their hard work in implementing the designs.

RFERENCES

[1] MEXTRA, *Northwest LIS VLSI Design Tools Reference Manual, Release 3.1*, NW Laboratory for Integrated Systems, University of Washington (1987)

[2] Dingwell, A. and Zazzu, V., 'An 8MHz 8b CMOS subranging ADC', *IEEE International Solid-State Conference* (1985)

[3] Baugh, C. and Wooley, B., 'A two's complement parallel array multiplication algorithm', *IEEE Transactions on Computers*, **C-22**, No. 12 (Dec., 1973)

ABSTRACTS–ENGLISH, FRENCH, GERMAN, SPANISH

Student full custom design

Final year students have been designing full custom integrated circuits for their final year projects, submitting their designs for fabrication and testing the packaged devices on their return. This paper describes the tools used, the designs undertaken and the benefits of allowing students to carry out full custom design projects.

Utilisation par des étudiants du matériel ECAD pour la conception de circuits dédicacés

Des étudiants de dernière année ont conçu des circuits intégrés dédicacés pour leur projet de fin d'études, soumettant leur réalisation à la fabrication et vérifiant les circuits encapsulés à leur retour. Cet article décrit les outils utilisés, les projets entrepris et les avantages obtenus en permettant aux étudiants de réaliser complètement des projets de circuits dédicacés.

Studentischer Vollkunden-Entwurf unter Nutzung von ECAD-Hardware

Studenten im letzten Jahr haben für ihr Abschlußprojekt Vollkunden-Schaltkreise entworfen und übergaben die Entwürfe für eine Fertigung, wobei nach Rücklauf der verkapselten Bauelemente eine Testung erfolgte. Dieser Artikel beschreibt die benutzten Tools, die erarbeiteten Entwürfe und den Nutzen für die Studenten durch die Ausführung von Vollkunden-Entwurfsprojekten.

Estudio del diseño full custom

Los estudiantes de ultimo año han estado diseñando circuitos integrados full custom para sus proyectos de fin de carrera, mandando sus diseños a fabricar y controlando los fallos una vez recibidos. Este articulo describe las herramientas utilizadas, los diseños realizados y las ventajas que supone permitir a los estudiantes abordar proyectos de diseño full custom.

A SOLUTION TO THE PIN-OUT PROBLEM IN MULTIPROJECT CHIPS

J. G. SWANSON and J. P. TILSTON
Department of Electronic and Electrical Engineering, King's College London, England

INTRODUCTION

Increasingly, the design of silicon integrated circuits is part of the training of electronic engineers and computer scientists. The combination of several designs on one chip and several chips on one wafer has lowered the cost of manufacture. Design, manufacture and testing can be accomplished within an academic year[1].

Unfortunately, placing several designs on one chip immediately reduces the number of pin connections which can be given to each design. The sharing of 40 pins between 4 students indicates that the individual designs need to be carefully selected. In a CMOS technology where care must be taken to avoid latch-up it is sensible to allocate a power supply connection to each design, further reducing the number of signal pins per design. The approach to be described here uses a multiplexed signal bus to permit the use of all of the remaining pins by each design.

AIMS

An important requirement is to be able to offer the transfer of binary words into and out of a design without using serial methods. A minimum of 16 pins would be needed if 8-bit bytes were used and in addition a number of clock and control pins would ne needed. The inclusion of power supply connections raises the total to at least 20. At this stage it is possible to contain two carefully chosen designs in one 40-pin package. There is, however, no possibility of using extra pins for testing and fault diagnosis. This aspect is as important to the teaching process as it is for ensuring testability.

The CMOS technology was chosen because of its industrial relevance. It was decided to use a separate power supply connection for each sub-design to prevent latch-up in one spoiling the others. Four designs were to be accommodated. A common ground was to be used for these designs and the chip frame. In this case, however, the latter was split and used two power supply and ground connections of its own. A further pin was reserved by the supplier, European Silicon Structures, to isolate the frame pads from the designs. Thus a total of 9 pins were already allocated, leaving 31 to be multiplexed. The aim, therefore, was to devise a scheme which made all of these available to four separate designs. It would have been possible, under different circumstances, not to split the frame and to dispose of the pad isolation line, thereby making a further 3 connections available.

Another aim was to avoid a rigid division of the design area, allowing large designs to be accommodated with smaller ones.

THE DESIGN

Signal bus
The use of a double metal layer CMOS technology offered topological flexibility in coping with crossing conductors. It also offered the use of CMOS transmission gates as signal switches. It was decided to use a common signal bus formed in metal-1 and to connect it via transmission gates using metal-2 and minimum area vias measuring 3×3 μm. This permitted a minimum bus conductor width of 5 μm and a separation of 3 μm.

The chip measured 3.28×2.65 mm and used a frame which contained the standard I/O circuits along the longer sides. The available design area within this was 3.05×1.69 mm. An approach, which was soon abandoned, was to quarter the available area with some of the signal bus lines running centrally north and south and the remainder centrally east and west. This was unneccessarily rigid and was replaced by a single set of 31 bus lines running centrally and perpendicular to the longer chip dimension i.e. north and south. The arrangement is shown in Fig. 1. The ends of the bus lines were turned through 90° to run along the edge of the frame to connect with the appropriate buffer

FIG. 1 An overall view of the bus meeting the chip frame.

circuit in the frame. Thus the available design area was reduced by the bus along its length and along the longer frame edges. This is most easily appreciated from an overall view of the chip in Fig. 3.

Connections could be positioned anywhere along the length of the bus and it was possible to flexibly position the two sub-designs in each 'half' of the chip. In principle, the bus need not be positioned centrally and could even contain a jog to accommodate different width designs in the two halves.

The bus had an overall width of 250 μm and provided 31 parallel signal lines each 5 μm wide and separated by 3 μm.

Access to the bus

The signal take-off lines, in metal-2, run east to west to transmission gates which isolate a design from the bus when it is not activated. Sixty-two transmission gates were needed on each side of the bus to provide two sets of 31 connections which could be activated separately. It proved difficult to fashion a unit which could be repeated 62 times without exceeding the available length, 1450 μm, in the bus direction.

A cell was designed which contained two transmission gates which were rotated versions of each other. This particularly compact pair, shown in Fig. 2, was repeated 31 times on each side of the bus. A disadvantage of this design was the need to serve each row of transmission gates with 3 control signals. The true control signal TX was fed to each cell on two further bus lines and one

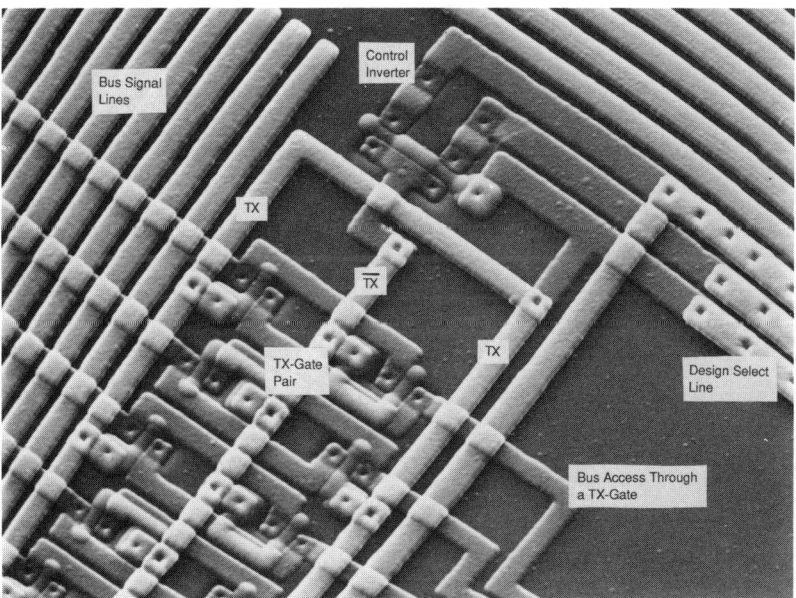

FIG. 2   Part of the bus with alternately rotated TX gates and their controlling inverter.

other was used for its inverse $\overline{\text{TX}}$. This arrangement was repeated on the other side of the bus so a total of 6 lines were incorporated to control the access switches.

Separate selection of the four designs was arranged by splitting each of these half way along their length and by feeding the independent control signals from the outer ends. These signals were supplied by four logic inverters, one for each design. The input of each inverter was simply taken from the power pin of the sub-design which was to be selected. So, on powering up a sub-design, the transmission gates from it to the bus were automatically conducting and the three other designs were isolated and without power; their power pins were grounded. Thus selection of a design was accomplished without devoting extra package pins to this function and without address decoding. Fig. 2 shows a scanning electron micrograph of part of the bus together with some of the transmission gates and their controlling inverter.

PERFORMANCE

Design selection
The implementation of this scheme formed part of a final year undergraduate project. Fabrication was carried out using a 2 μm technology by European Silicon Structures (ES2) on a multiproject wafer arranged by the University of

FIG. 3   An overall view of the chip showing the bus and the four design areas.

London VLSI Consortium. The design exercise including testing was completed within an academic year.

The chip accommodated a number of circuits, unconnected with this scheme, arranged in the four sub-design areas. Fig. 3 shows an overall view of the chip. Each of these designs could be activated and isolated from the others by selecting the appropriate power supply pin. It was interesting, and perhaps useful for the future, to note that the activation of design areas could be cycled at frequencies up to 500 KHz.

Signal delays

An assessment of the average propagation delays of a number of simple logic gates was made by combining them in odd numbers into ring oscillator circuits. The frequency of the oscillators provided a measurement of delay per stage which was not affected by delays in the frame buffer or the signal bus. A typical delay for a minimum geometry inverter was 10 ns. Whenever a signal was routed out through the multiplexed bus and frame buffer a delay of 10 ns was added. There was no noticeable improvement when the TX gate was omitted from the route. The 10 ns delay was close to the manufacturers value for the pad buffer, so it can be concluded that the bus and transmission gate introduced negligible delays. The maximum operating frequency was 50 MHz, this was set by the frame buffer.

Cross-talk

The particular set of connections to the bus was not specifically organised to assess cross-talk between lines. With no sub-designs activated test signals were applied to the bus lines through the pad I/O circuits. When a 5 V, 50 KHz square wave was applied the interference on an adjacent line was less than 5 mV. A more realistic test investigated the coupling onto bus lines from the output of a ring oscillator running at 50 MHz within one of the sub-designs. The resulting signals did not correlate with line proximity in the bus but spikes as large as 1 V were evident. Although this was large it was well below the measured noise margins of the logic gates which were about 2.2 V.

CONCLUSIONS AND DISCUSSIONS

The viability of using a multiplexed signal bus to allow several sub-designs to have access to all of the non-power supply package connections has been demonstrated. Design selection by using individual power supply pins to power up the designs and to give access to the bus works well.

In this implementation, 4 sub-designs, each used 31 signal pins on a 40 pin package. The design area consumed by the bus and selector switches was 24% of the total available. The two regions separated by the bus measured 1.3 × 1.5 mms. A further 3 signal pins could have been used if a split frame had not been used and the frame isolating line had been abandoned.

Further investigation of cross-talk would be desirable. A small sacrifice in design area would allow the bus line spacing to be increased and this, together with the elimination of redundant lengths of bus line, would be helpful in reducing bus-line coupling.

Great care must be taken, at the layout stage, to avoid confusing lines when the connections are being arranged to the bus. A labelling scheme at this stage would be an asset.

The scheme produces dramatic reductions in cost per student when pad limited designs are to be made. This is frequently the case when extra connections need to be available for proving elements of a design in an educational setting. The scheme is extendable, in principle, to accommodate more designs per chip at the expense of bus area.

ACKNOWLEDGEMENTS
I would like to express my thanks to Sandy Davidson and Alan Kent, respectively Coordinator of the University of London VLSI Consortium and the University of London Industry Training Partnership, for critically commenting on the manuscript.

REFERENCES
[1]   Swanson, J. G., 'VLSI design at the undergraduate level', *Int. J. Elect. Engng. Educ.*, **24**, pp. 309–318 (1987)

ABSTRACTS–ENGLISH, FRENCH, GERMAN, SPANISH

A solution to the pin-out problem in multiproject chips
Combining several relatively small student designs on one chip awkwardly reduces the number of external connections per design. A multiplexing scheme is described which allows four students to each use 31 pins on a 40-pin package. The implentation of this scheme in a 2 μm full-custom CMOS technology is reported.

Une solution au problème du brochage dans des puces multi-projets
La combinaison de plusieurs projets relativement réduits d'étudiants sur une seule puce rëduit de façon embarassante le nombre de connexions extérieures de chacun des projets. Le schéma de multiplexage décrit permet à quatre étudiants d'utiliser chacun 31 broches sur un conditionnement à 40 broches. L'implantation de ce schéma en technologie CMOS 2 μm dédicacé est signalée.

Eine Lösung des Pinproblems in Multiprojekt-Chips
Die Kombination mehrerer relativ kleiner studentischer Entwürfe auf einem Chip reduziert die Zahl externer Verbindungen je Entwurf in unvorteilhafter Weise. Eine Multiplex-Schaltung wird beschrieben, die es vier Studenten ermöglicht, 31 Pins auf einem 40 Pin-Gehäuse zu nutzen. Über die Implementation dieser Schaltung in einer 2 μm-Vollkunden CMOS-Technologie wird berichtet.

Une solución para el problema de Pin-Out en multiproyectos chips
Combinando varies proyectos de estudiantes en un chip dificilmente se reduce el número de conexiones al exterior. Se describe un sistema multiplexor que permite a cuatro estudiantes usar a cada uno 31 pins de un total de 40. La implementación de este esquema se ha realizado en una tecnologia CMOS full custom de 2 μm.

THE USE OF ELLA IN TEACHING AND RESEARCH

A. J. PEARMAIN, A. R. RAZI and P. R. COWARD
Department of Electrical and Electronic Engineering, Queen Mary College, University of London, England

INTRODUCTION

ELLA* or the *Electronic Logic Language* was initially developed by a small group at the Royal Signals and Radar Establishment (RSRE), starting in 1978, as a design language, compiler, design management environment and simulator for the design of relatively complex digital systems. The marketing of the software was taken over by *Praxis Systems plc* in 1985 and there is continual development of the system.

The objective of ELLA is to allow design to be undertaken at various levels of detail, from the highest level of abstraction down to logic gate level, and to enable checks to be made of the consistency of the design between the different levels of detail. The simulator allows circuits described in the ELLA hardware design and description language (HDDL) to be simulated and support is provided for parallel processes to occur in the circuit. It has now been adopted very widely in UK industry, and to a lesser extent in other countries, as part of the integrated circuit design process.

The U.S. Department of Defense (DOD) also identified the need for such a language and set up a committee to design a standard language that would become mandatory for IC design under DOD contracts. This language is known as VHDL and is an obvious competitor for ELLA. However, whilst the standard for VHDL has been available for some time, simulators are only now appearing. In a teaching environment, the richness of syntax in both languages causes problems, but the problem is much greater with VHDL which aims to be more general than ELLA.

ELLA was distributed to members of the Electronics Computer Aided Design (ECAD) initiative in the UK in a VAX† version in summer 1987, with the Apollo version being distributed in summer 1988.

---

*ELLA is a registered trademark of the Secretary of State for Defence.
†VAX is a registered trademark of Digital Equipment Corporation.

GENERAL PROBLEMS OF USING ELLA IN AN EDUCATIONAL ENVIRONMENT

The general problems encountered are:

- *Disc space* If a design of significant complexity is described in ELLA and compiled into the database, the file is relatively large. This is not important when the whole of the disc space on a workstation can be used, but the default space allocated by the computer centre per student on the Queen Mary College (QMC) VAX system (150 k) was rapidly exceeded and students had to be allocated 500 k per student to enable work to be done.

- *Learning the language* The language is necessarily complex to enable many different structures to be correctly described. The language syntax is also distinctly different from the languages with which our students are familiar (Pascal and Lisp). There is therefore a distinct resistance to using the language.

- *CPU time* Simulating a circuit using ELLA takes a considerable amount of CPU time. This causes a problem on a multi-user system, or on a workstation if attempts are made to run other tasks simultaneously.

ELLA IN TEACHING

We are using ELLA in the course *VLSI Design* that is a final-year option, but one that is recommended for the programmes of study being followed by about half out students. It has a value of one course unit, which means that it is supposed to represent 25% of the final year work load. The history of this course, which has been run since the 1984/5 session with 3 hours of classroom time per week for 22 teaching weeks per year, has been described elsewhere[1,2]. There have been changes in the course organisation and content each year in the light of developments in the subject and in the equipment available. Our priorities in the subject matter covered have changed considerably over the four years. Initially we included a significant amount of background physics of semiconductors in the course, together with basic NMOS structures and simulation techniques, but we have now reduced the physics content, concentrated on CMOS structures, brought in much more material about CAD techniques, design for testability and testability analysis, and decided to look in much greater detail at chip architecture and the management of complexity. It is in this area of architecture and design management that ELLA has its role.

Students have always undertaken two IC design projects as part of the course, with 2 hours per week of workstation time scheduled for each student. The design activity accounts for 33% of the marks for the course. The first design exercise involves using a UNIX* workstation and some full-custom IC design tools to design a relatively simple CMOS circuit (a variation on an 8-bit counter in 1987/8), selected examples of which are fabricated and returned for testing. Students work in pairs on this exercise.

---

*UNIX is a registered trademark of AT&T.

The second design exercise aims at looking at the design process at a higher level of abstraction. In this exercise students work in groups of 4 or 5, partitioning the design between them. So that a design can be more complex, PLA-generator tools are used to avoid the detail of the layout process. ELLA was introduced into this second design exercise in the 1987/8 session. The design exercises that were undertaken were:

- Floating point adder/subtracter, 16 bit bus compatible, 32 bit internal.
- IEEE 488 to RS 232 interface.
- Floating point multiplier.
- UART.
- CPU using hard-wired controller.
- 4-digit, 7-segment counter display driver.

These cover a considerable range of complexity, with some being much too ambitious for complete implementation in the time available.

The intention was that the group should specify the design and its partitioning at a high level of abstraction initially, which would be written in ELLA and simulated. The whole class had three lectures about ELLA, backed-up with a 9 page document that we had prepared to define a very minimum sub-set of the language. One person from each of the six design groups was selected to undertake the ELLA part of the project.

In the one year that we have run this part of the exercise only 50% of the groups succeeded in completing the ELLA description and simulation of the design. In part this was due to difficulties with using a different computer system from the one that they were using for the rest of their design work, rather than the language. Unfortunately ELLA was only available running on the VAX machine under the VMS operating system, which required the students to use an unfamiliar operating system as well as an unfamiliar language. This machine is also under the control of the computer centre and is heavily used by other courses such as introductory programming courses and courses using statistical analysis packages. The access that students had to the machine was therefore quite restricted. We also had to learn by trial and error that students required a larger filestore allocation than we had anticipated.

An additional problem is the competition for student time between the coursework for this course, coursework for programming courses and final-year projects. There is a distinct danger that students will be overburdened with coursework with the pressures to satisfy professional institution requirements for Engineering Applications, EA1 and EA2 education. We also believe that students do not obtain the same immediate stimulus from working with a tool that operates entirely in text mode on dumb terminals as they do from working with colour graphics interfaces, so the students give the ELLA work a low priority.

We are firmly committed to the principle of using ELLA as part of our VLSI Design course. The issues of architecture and design management are very important within the subject and have strong parallels in the topic of software engineering. We are pleased that some of our students were successful with using ELLA for the top level of their design process, but disappointed that this was only a small proportion of the class (effectively three or four students out of a class of 25). We shall discuss our plans for improving the situation below.

ELLA was also used in a final-year project by a student who was sponsored by an employer that uses ELLA regularly as a tool in their design process. This proved to be a successful project, with the student bringing some expertise from his sponsor that was useful to us in increasing our understanding of the language.

FUTURE TEACHING PLANS

We intend to repeat the inclusion of an ELLA description and simulation as part of our second design exercise. This exercise will use the Solo 1200 silicon compiler as the layout tool in the 1988/89 session and both this and ELLA are now running on our Apollo system. This will considerably reduce the problems of using the computer system and provide a much better capability of integrating the ELLA part of the work with the layout. The provision of a translator from ELLA to the *Model* HDDL (the language used by the Solo software), would permit a complete integration of the design process, although the design philosophy of ELLA is somewhat different from that of Model, which might make it difficult to provide such a translator. Use of Solo will reduce the amount of effort that students have to expend to complete an architecturally interesting design and it will control the design process in a more formal manner that is more typical of a commercial design environment.

We also intend to improve our introductory documentation for ELLA and to prepare a guided *walk-through* exercise that all students would perform. This will ensure that all students gain a better understanding of ELLA and will be good preparation for those in each design team who will be responsible for the ELLA description and simulation part of the design.

ELLA IN RESEARCH

Among other research activities in the department of Electrical and Electronic Engineering at QMC, we are in the process of developing a systolic array chip for signal processing purposes in adaptive radar systems.

An adaptive processor is the heart of any adaptive array system. It automatically adjusts the weights on incoming signals to achieve the desired spatial filtering. The adaptation process is based on an algorithm, the one selected in this case being the Gram-Schmidt algorithm. This algorithm is partly selected because it can be mapped into hardware as a number of identical building blocks or 'cells'. This is ideal for implementation as a systolic adaptive processor.

Our first task was to select the best chip architecture for implementing the algorithm as a chip or set of chips. The only way of evaluating and verifying our designs against the specification prior to fabrication is by simulation, but it is important that we should be able to test variations on the architecture of the chip without detail design of each variation of architecture. However, it is impossible to obtain realistic timings for high-level blocks without some detail gate design.

Because ELLA is intended to describe electronic hardware, it has a hardware model that is related to the structure and behaviour of real circuits. The ELLA simulator can handle any type and value of signal that has been defined and enumerated by using TYPE declarations. The simulation performance is faster and more efficient because the simulator is driven by a fully-compiled code. The simulation times are drastically reduced compared with conventional gate-level simulators.

The language has been designed so that it is easy to manipulate repeated elements in hardware. As our design is a series of identical blocks of processors connected in horizontal and vertical fashion, the command MACRO was used to manipulate repeated 'cells'. In this way an array of n by n cells could easily be described.

There are two ways of describing such repetitive hardware using the MACRO. The first uses an iterative construct, which replicates the hardware objects that follow it. The second, more useful and more powerful, is recursion. Recursive macros were used in our work to produce a triangular structure. An advantage of using the iterative construct is that errors are more easily detected, because any error will be multiplied so that the fault will be obvious when the circuit is simulated.

Because ELLA describes a circuit as a network of nodes and subnodes, connected by wires, it allows complex circuits to be described in a hierarchical manner. This is very useful at the elementary stages of the design, since it lets the circuit designer concentrate on the design as a whole, with the internal details being added later.

In some ways ELLA is a relatively small language, and there are only four basic ways of describing the behaviour of a node in a network. Two of these behavioural primitives, CASE and ARITH, transform the values of signals passing through a node. CASE is used to describe a truth table and ARITH is used to describe circuits that process numerical signals without describing the hardware that executes the arithmetic.

The main advantage of the ARITH statement is that large simulations can be speeded up, as the functionality of a component is described by arithmetic rather than by a complex network and it is this function that is most powerful in allowing outline design at a high level of abstraction.

At present our main concern is the design of the internal circuits of the 'cells' in our processor. This will be a fairly complex design as a large number of multiplications and additions has to be performed. Progress was again frustrated by lack of access to the system, which was exacerbated by a policy that teaching has priority over research on the VAX on which ELLA has been mounted. This frustration has now ended as we now have ELLA mounted on our Apollo system.

Our initial reaction to attempting to use ELLA in our research programme is that it will be invaluable as a technique for rapid outlining of architectural solutions to a problem, with the capability of then ensuring that the detailed design is consistent with the original specification and partitioning.

We would be very excited about a link between ELLA and Solo because this would enable almost all the effort to go into problems of architectural innovation instead of layout detail, with the promise of a very low risk of failure when designs were fabricated.

CONCLUSION
The inclusion of ELLA in the ECAD initiative allows students to experiment with chip architectures and design management in a way that would not otherwise be possible. Initial experience suggests that it is quite possible to include work with ELLA in the final year of a three-year undergraduate programme, but that this will require a carefully prepared introductory exercise to be completely successful.

ELLA is a very powerful tool for use in research. It allows relatively rapid outline simulation of chip architectures so that a number of possibilities can be evaluated. It will prove to be of even greater value if it can be combined with a silicon compiler to give a high-confidence rapid route from architecture to silicon.

The major problems that have been encountered in using ELLA have been the availability of adequate computing resources for the requirements of ELLA and resistance to having to learn yet another language.

REFERENCES
[1]   Pearmain, A. J., 'Using Metheus workstations for student VLSI design at Queen Mary College', *Silicon Design*, **2**, no. 3, pp. 5 & 7 (1985)

[2]   Pearmain, A. J., 'Teaching silicon circuit design in a degree course', *University Computing*, **8**, no. 2, pp. 109–112 (1986)

ABSTRACTS–ENGLISH, FRENCH, GERMAN, SPANISH

The use of ELLA in teaching and research
The ELLA hardware description language and simulator are used at Queen Mary College in a final-year course where students simulate the specification of a design to verify the correctness of the specification and the definitions of the design partitions. Research use is in evaluating alternative parallel architectures for signal processing.

Utilisation de ELLA dans l'enseignement et la recherche
Le langage de description de matériel et le simulateur du système ELLA sont utilisés au Queen Mary College dans un cours de dernière année où les étudiants simulent la spécification d'un projet pour vérifier l'exactitude de la spécification et les définitions des partitions du projet. L'utilisation en recherche consiste en l'évaluation de différentes architectures parallèles pour le traitement de signaux.

Die Nutzung von ELLA in Lehre und Forschung
Die ELLA Hardware-Description-Language und der Simulator werden am Queen Mary College in einem Abschlußjahr-Kurs genutzt, in dem Studenten die Spezifikation eines Entwurfs simulieren,

um die Richtigkeit der Spezifikation und die Definitionen der Entwurfsteile zu verifizieren. Die Nutzung für die Forschung besteht in der Entwicklung alternativer paralleler Architekturen für Signalprozessoren.

El uso de ELLA en investigación y docencia

La descripción del languaje y simulador del hardware ELLA se ha utilizado en el último curso del Queen Mary College donde los estudiantes simulan la especificación de un diseño para verificar la previsión de la especificación y las definiciones de las particiones del diseño. En investigación se utiliza para evaluar arquietecturas paralelas alternativas para procesadores de señal.

USING ELLA* AS A DESIGN TOOL

P. R. MILLER, M. ZWOLINSKI and C. R. JESSHOPE
Department of Electronics and Computer Science, University of Southampton, England

1 INTRODUCTION

The advancement in VLSI fabrication techniques, coupled with the nature of existing design tools and the pragmatism of design methodologies has led to a dilemma of complexity in chip design making it a major bottleneck in production. The labour intensity required questions the economic feasability, with the implications being that many potential VLSI applications remain unexplored. Indeed, this gap is analogous to the *software lag* relative to the capacity of computing hardware. As design complexity grows, another major concern unfolds. That is, one of correctness. With product life-cycles shortening, the concept of *right-first-time* becomes simultaneously more important and more difficult to achieve. Whilst logic simulation must be an integral part of any chip design, it cannot alone deal with the complex problems of evolving a correct specification and developing a chip to meet it. A behavioural description, coupled with a solid design methodology provides an approach that addresses the issues raised.

ELLA is one example of a behavioural description language or HDL (Hardware Description Language). Logic or gate level descriptions, as well as being time-consuming and detailed, emphasise how the design works, but say very little about what it does. In contrast, ELLA provides a means of specifying what a design does, without necessarily describing how it does it. Behavioural modelling relieves the burden of implementation details, thus placing the emphasis on specification. However, since this specification can be animated, and in much shorter times than via a gate level simulation, the exploration of design alternatives now becomes a viable proposition. Since it is possible to manipulate configurations, design partitioning is eased, which in turn leads to a mixed-level development of designs. Decisions can be deferred on various components, leaving them in a black-box state, whilst other modules can be fully developed. In short, it allows the designer to dissolve the problem into both conceptual and manageable blocks.

An overview of the ELLA system is given, followed by a description of how it has been used as an architectural design tool to develop packet-switching

---

*ELLA is a registered trademark of the Secretary of State for Defence.

intercommunication networks for SIMD processor arrays. Finally, with the experience gained from this work, a discussion of the merits and drawbacks of ELLA is presented, together with a summary on current trends in automated design and ELLA's role in it.

2   AN OVERVIEW OF THE ELLA SYSTEM

Ella[1] is an integrated hardware design system consisting of the ELLA language and its compiler, the ELLA Application Support Environment (EASE), a simulator, and ELLANET.

Ideas, abstract designs and detailed designs are described using the ELLA language. As with other Hardware Description Languages (HDLs) the approach is a departure from conventional von Neumann programming methodology. Objects, or more precisely FUNCTIONS, are created with explicit declarations of inputs and outputs. The ELLA text of each function describes its behaviour at whichever level of abstraction the designer chooses, and then functions are simply connected via their inputs and outputs. At simulation time, the network of functions and interconnections work in pseudo parallel, thus descriptions in ELLA map onto real hardware concepts and the designer is placed in a familiar world.

Descriptions can be partitioned and integrated within EASE, which is constructed on a purpose-built database. This manages the design data, allowing hierarchical and multi-level descriptions to be built up, and provides a facility for the extraction of data for use by ELLANET. Animations of design descriptions take place in the simulation environment, where network inputs can be changed and network nodes monitored and displayed. ELLANET provides a medium within which programs can be written to use the data provided by EASE to interface to other CAD systems.

The ELLA language

The language is both simple to use and easy to learn. Rather than providing a multitude of specific facilities, it has only a few general and comprehensive primitives. As ever, an example serves best to illustrate the basic principles of description.

Consider first the description of an RS flip/flop constructed from cross-coupled NOR gates. The circuit schematic is shown in Fig. 1.

```
Type boolean = NEW (true|false)
FN NOR = (boolean:ip1 ip2) - > boolean:
CASE (ip1,ip2)
OF    (false,false):true,
ELSE false
ESAC.
FN RSFLIPFLOP = (boolean:set reset) - > boolean:
```

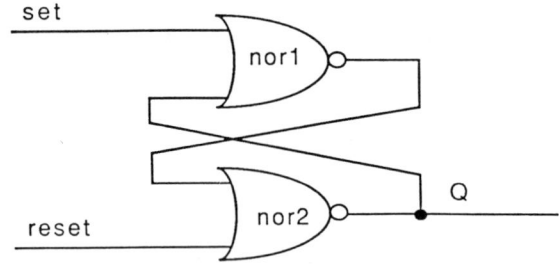

set

nor1

reset

nor2

Q

FIG. 1   RS flip/flop schematic.

```
BEGIN
      MAKE NOR:nor1 nor2.
      JOIN  (set,nor2) - > nor1,
            (reset,nor1) - > nor2.
      OUTPUT nor2
END.
```

Initially the signal types to be used in the network must be defined, and in this example we declare a simple boolean variable to be used throughout. Next the NOR gate element is described, having two inputs and a single output. The CASE primitive is used to describe the required output given the appropriate inputs, with the output of the CASE clause being the output of the function. Since this is the only construct of the function, no BEGIN and END are necessary, and no explicit OUTPUT must be declared as there can be no ambiguity.

The RSFLIPFLOP function MAKEs two function instances of the NOR gate, nor1 and nor2, and then simply JOINs them in the required configuration. In this situation two nodes are created, so the required output must be stated explicitly.

This was an example of a low-level description. Since the component is rather simplistic there would be little to gain from abstracting the description, but in a more realistic application the advantages become clear. Suppose it is required to use the RSFLIPFLOP function as a start/stop circuit for an eight-bit counter. The counter could of course be described at gate-level, but that would be both time-consuming and a burden when considering simulation times. Alternatively, the counter could be described behaviourally at this early stage, and then fully designed nearer system completion.

Before commencing with the description, the nature of simulation, or more precisely the concept of time in the simulator, must be understood. Combinatorial logic, without delays, will always evaluate regardless of the advancement of the simulator time-base. However, memory elements, created using the RAM and DELAY primitives, will only update upon the specified number of time increments. It is of course possible to omit the use of the time-base from designs, but to achieve this with a design containing memory elements would involve full logic descriptions. When abstractions are required, substituting the system clock with the simulator time-base is ideal.

The abstract counter function can now be described. Figure 2 shows a diagrammatic representation of the text.

```
TYPE int = NEW i/(0..255).
FN COUNTER = (boolean:enable clear) - > int:
BEGIN
        FN DELAY_INT = (int) - > int:DELAY(i/0,1).
        FN INC_INT = (int:ip) - > int:
        ARITH IF ip = 255 THEN 0 ELSE ip + 1 FI.
        MAKE DELAY_INT:del,
              INC_INT:inc.
        JOIN  CASE clear
              OF    true: i/0,
                    false: CASE enable
                          OF    true:inc,
                                false:del
                          ESAC
              ESAC- > del,
              del- > inc.
        OUTPUT del
END.
```

The new signal type 'int' is a further design abstraction. Rather than deal with eight boolean outputs an ELLA integer can be used. The function INC_INT

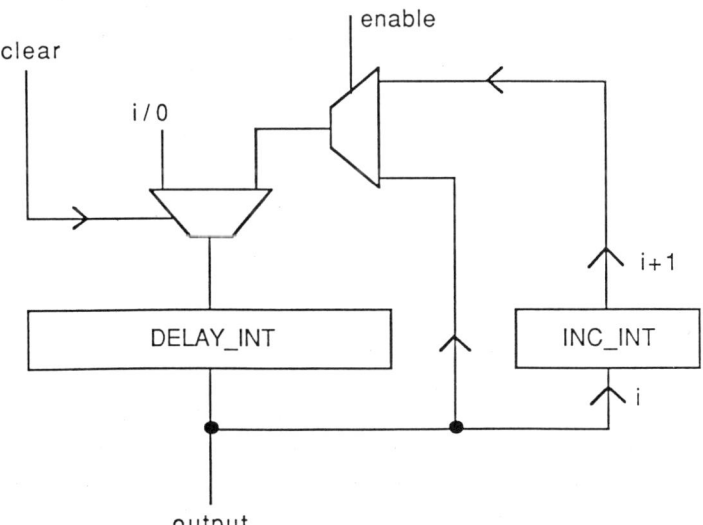

FIG. 2   Function COUNTER schematic.

utilises the ARITH primitive which can use *if...then* constructs and manipulate ELLA integers with a full range of arithmetic and logical operators. Bit-wise logical operators are also possible upon integers.

Another point to note here is the way in which primitives can be nested. The function instance 'del' is either joined to itself, thus retaining its value through time, or joined to the 'inc' function, thus having its present value incremented. This is achieved by nesting the case clause within the JOIN statement.

Having now defined the constituent parts of the example design, the top-level module can be constructed:

```
FN STOPWATCH = (boolean:start stop clear) - > int:
BEGIN
        MAKE RSFLIPFLOP:rsflipflop,
              COUNTER:counter.
        JOIN   (start,stop) - > rsflipflop,
              (rsflipflop,clear) - > counter.
        OUTPUT counter
END.
```

It is neither appropriate nor is there room here to discuss all the features of the ELLA language. However, some are worthy of mention. Associated types for signals are possible, which introduce a hierarchy of signals on one connection such as

defined&data|undefined

Of the memory elements only DELAY has been introduced. There also exists a RAM primitive having address, data, and read/write variables. Parameterised functions can be created by using the macro command MAC, to achieve such functions as n-bit adders. The use of this in an ELLA component library is obvious. Function types allow the user to describe bi-directional signals as a single ELLA objects. This is a novel feature not found in other HDLs.

The example shown serves as a suitable introduction to the basic concepts. It has demonstrated the central features of the language and hopefully illustrated both its power and flexibility. As other features of the ELLA package are described the STOPWATCH example will frequently referred to and expanded upon.

The ELLA application support environment (EASE)
EASE is the environment into which designs are embedded prior to simulation. A system of contexts (work areas) can be created, each independent of the other, within which units of ELLA text can be compiled and linked. Sub-modules of an architecture can be developed within their own contexts, with a hierarchy of contexts structured to reference other contexts thus consolidating the design at the top-level. Furthermore, background utilities continually check design consistencies, such as matching signal types on interconnects, giving

details of errors and preventing simulation until the fault is resolved. This is an extremely useful and essential aid to design development ensuring consistency throughout all levels of the hierarchy. EASE is constructed on a purpose-built database system that transparently manages data on disc. Therefore the user needs to know little about the host operating system to work within the ELLA environment.

Compilation is performed within EASE. Incremental mechanisms allow functions to be compiled separately and then added to those already stored within the working context. Furthermore, updates simply overwrite previous declarations. To compile successfully, as well as being syntactically correct, declarations must be consistent with those already in the database. Thus, recompiling a new version of a function will require any declarations that reference that function to also be recompiled. Again, consistency in design is ensured. The compiler is reasonably fast (2000 lines/minute(cpu) on DEC VAX 11/780), and produces concise, direct diagnostic information which in conjunction with the relatively orthogonal language provides a short and mainly trouble free development time.

The context system is used to partition design data, normally to reflect the hardware structure being described. Of course, declarations can be explicitly shared in a controlled manner. As a rather trivial example of this, suppose FN RSFLIPFLOP has been compiled in a context called 'flipflops', which could contain a library of such devices. Now, if FN COUNTER and FN STOPWATCH are to be compiled within a context called 'countdevices', two operations, namely IMPORT and EXPORT must be performed. Firstly, from within 'flipflops' the following command line must be entered (note that the system prompt becomes the current working context name):

```
flipflops <-export RSFLIPFLOP <ret>
```

This has the effect of making the function globally accessible for import by any other context. It can be cancelled by the counterpart command 'dontexport'.

Within 'countdevices', and in the same source file containing COUNTER and STOPWATCH, the following line of code must be added to access the required RSFLIPFLOP function.

```
FN RSFLIPFLOP = (boolean,boolean) - > boolean:IMPORT.
```

This will generate an empty function cell within 'countdevices' with the specific input and output types. To fill the shell, the following text file must be compiled within the same context:

```
IMPORTS flipflops:RSFLIPFLOP.
```

The closure of 'countdevices' is now complete and simulation can commence.

The system for importing and exporting ELLA declarations between contexts provides the user with a sophisticated system for performing multi-level simulations. Alternative descriptions of components can be imported into the simulation context at different levels of abstraction.

There is an extensive EASE command language supporting a number of utilities for manipulating designs and checking design consistency. These can be typed interactively or read from command files.

The ELLA simulator

The ELLA simulator can be used from the development of system architectures down to gate-level validation, therefore being ideally suited to a top-down design methodology. Simulation times can be dramatically reduced by using this mixed-level capability. Well defined and tested features of a design can simply be described behaviourally and therefore be less demanding timewise. The more novel aspects of the design can be developed progressively right down to gate-level, with cross-referencing ensuring system behaviour is unaltered.

There are a number of ways of inputting data to the simulator. Parameters can be typed interactively at the keyboard or a sequence read from a file. Alternatively, and more preferably with large systems, a test harness can be described and applied to the system inputs.

Returning to the example, the function STOPWATCH has three inputs, start, stop and clear respectively, and a single count output. Once in the simulator the following command lines can be applied to demonstrate the type of interaction one receives.

```
sim <-cp false true true,ti + 1 <ret>
        STOPWATCH = i/0
        **time = 1**
        STOPWATCH = i/0
sim <-cp false false false,ti + 1 <ret>
        STOPWATCH = i/0
        **time = 2**
        STOPWATCH = i/0
sim <-cp true false false,ti + 1 <ret>
        STOPWATCH = i/0
        **time = 3**
        STOPWATCH = i/1
sim <-cp false false false,ti + 1 <ret>
        STOPWATCH = i/1
        **time = 4**
        STOPWATCH = i/2
........etc
```

The command line is composed of the command *changeparameters*, abbreviated to cp, followed by the required input parameters to the system, given in the order in which they were declared. 'ti +' advances the simulator time-base by the specified amount.

There are three types of output display from the simulator: freeform text, tabulated text and graphics. The above example shows the default freeform text, with the only output shown being that of the function being simulated. A whole suite of display and monitor commands are also available so that any node may be displayed either continually, or only when it changes state. Tabular form is also possible with the table layout being automated, as well as a graphical interpretation of output signals on a suitable monitor.

Ellanet

Ellanet is an interface between the EASE database and other CAD systems. It provides the designer with the ability, through written programs, to access database information and transfer it to another environment. Several different levels can be accessed but a typical output would be a netlist.

Pascal or Algol 68 can be used to access the data structures contained in the EASE database. This is supported by procedures and identifiers provided for the task. The information extracted can then be manipulated into whatever form is necessary for input to lower level tools, such as gate array placement and routing systems.

3 ELLA AS AN ARCHITECTURAL DESIGN TOOL

The concept of a design discipline, as bounded by the applied range of mental and physical processes, is forever undergoing redefinition and expansion. The tendency is to reason within this limiting framework to which a design must conform. A transitional period in the discipline of computer architectures is now apparent, being a consequence of both advancements in VLSI capabilities and developments in automated design tools. The cost of processing logic is no longer an economic hurdle, opening areas of design freedom to the computer architect, and initiating a departure from the von Neumann paradigm[2]. Parallelism, in its many guises, offers the designer the ability to approach problems that were once computationally intractable by implementing architectures in which a large number of processors cooperate to solve a single task.

Aside from the fundamental issue of *multiple instruction multiple data (MIMD)* or *single instruction multiple data (SIMD)* operation there is growing awareness that data communications is the key to successful exploitation of parallelism. The ability of any single processor to communicate as efficiently as possible with any other in the system is of paramount importance. The advent of VLSI, coupled with the heightened awareness of effective communications has provoked a great deal of research into interconnection strategies.

Much work has been carried out recently on parallel architectures at Southampton University. This includes the full hardware development of the RPA (Reconfigurable Processor Array)[3], which employed a circuit switched nearest neighbour interconnect strategy, and a paper design, entitled μPA[4], borne out of the experience gained from the RPA project. This again utilised the nearest-neighbour square lattice topology, as studies have shown this to be cost-effectively the most suitable VLSI implementation. To enable full processor–processor communications across such a structure both hardware and software support were provided to effect an efficient packet-routing algorithm working concurrently with the processing elements.

To develop the communication processors towards completion of hardware and microcode algorithms ELLA was chosen. The reasons for this were two-fold. Firstly, the abstraction qualities of ELLA enabled the paper design to be effectively translated directly into code and animated without working down to an unnecessarily complex design early on. As the designer evaluated the architecture, the impact of design iterations and optimisations are consequently minimal. Secondly, with knowledge of envisaged sequential programming extensions to ELLA (version 3.0)[5], the development of microcode algorithms to control the hardware could be described more elegantly. It is worth noting that the partition between hardware and software has naturally evolved.

The eight-bit architecture of the packet-routing communications processor comprises an ALU, a register set, and a switchable RAM to enable data exchanges between the processing element and the communications processor. This interfaces to a set of datapath shift-registers which are used for trans-mission, distribution and reception of inter-PE information (Fig. 3).

The functionality is of little interest here. However, a brief discussion of the subsequent ELLA descriptions serves to illustrate its use in a more realistic application than the STOPWATCH example.

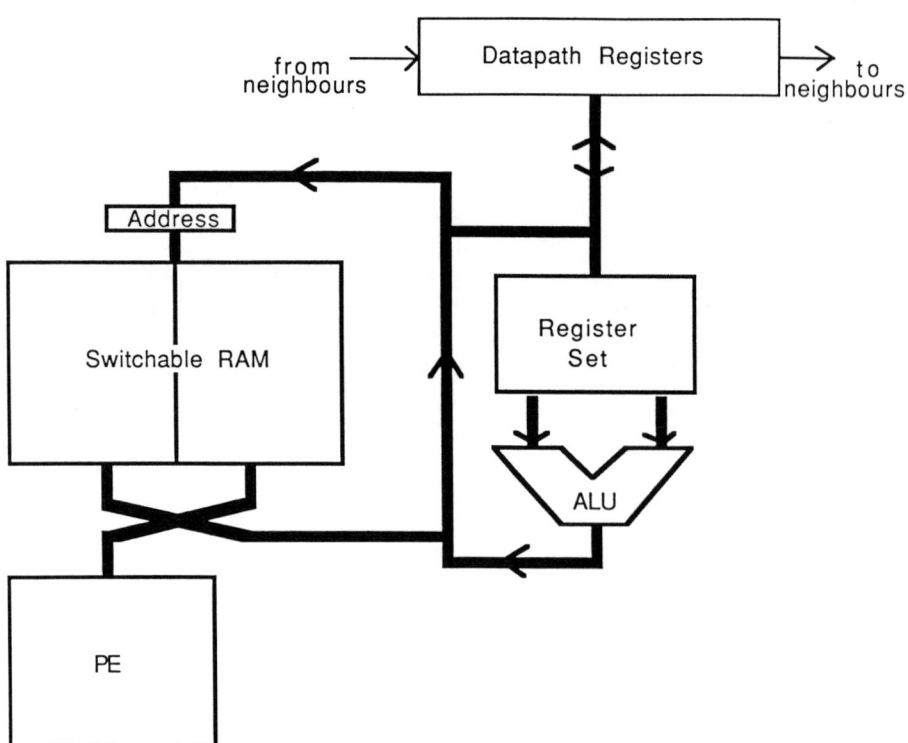

FIG. 3   Architecture of the communications processor.

The register set consists of eight identical eight-bit registers, each independently addressable. Using an abstract datatype of an ELLA integer, in the range 0..255, and mnemonics for the register addresses, the text required was minimal. The ALU description was trivial, again using integers to avoid bit-level complications. The ARITH primitive enabled the module to be described in only a few lines of text. Two instances of the RAM primitive, switchable by multiplexing (CASEing) the address, data and read/write lines, provided the buffer description.

The datapath shift-registers yielded the most complex description. To interface with the rest of the system the same integer datatype was used. The nine-bit shift-registers contain an extra *validity* bit to signify relevant data, so to simulate bit-level shifting a DELAY of nine time units was introduced. Intermediate values (between 0 and 9 shifts) were not required so such an abstraction presented no compromises. Once described, microcode development could commence, which simply coordinated the control signals to each module in the desired manner.

The process of formalising the architecture using ELLA and implementing microcode resulted in significant changes to the architecture from the initial paper design. Furthermore, the informal, hand-written microcode algorithms were found to be incorrect, the outcome being that correct versions were rewritten, and with the ability to realise the effect of these changes, optimisations resulted in a more efficient system than at first expected. This illustrates the importance of formalism in design.

4   EVALUATION OF ELLA

An invaluable part of the RPA project was the development of hardware simulation tools and interactive microcode editors[6]. It was not until the completion of these that various design problems were highlighted and corrected. However, the tools were implemented using a sequential programming language and in total amounted to about one year's man effort. In contrast, the ELLA simulation of the μPA communications network required only one month's effort, which included familiarisation with the ELLA system. It is probable that a full ELLA simulation of the μPA could be completed within another month. The implications are obvious. Design changes and optimisations can be detected and corrected at a much earlier stage.

Perhaps more important than any of the benefits gained from using ELLA in this application is the focus is provokes. The architecture discussed is now in a suitable form to be developed down to silicon, but it is unlikely to follow this path. The reason for this sudden halt is that work is now centred on the development of a new novel architecture with potentially far greater performance capabilities than its predecessor[7]. It is fair to say that the use of ELLA highlighted a number of inefficiences and so contributed, in part, to the new approach.

Avoiding repetition, it is felt that the advantages of using such a design tool have been presented. However there must surely be disadvantages too. In principle there are none, but in practice a few minor points are worthy of note. In abstracting a design the benefits are obvious. However, there arises situations where it can cause compatability problems, and as an example consider the loading of an eight-bit flag register from an eight-bit integer typed bus. Each bit of the loaded register needs to be decoded to be of use, whereas if the register were loaded from a true eight-bit bus then no adaptation would be necessary. The abstraction is still valid, but some overheads are incurred.

As a general complaint, the presentation of the software could be significantly improved, especially since versions are now available for the latest SUN and VAX workstations. All commands have to be typed, or read from a file, whereas a windows and menu-driven front end, with the ability to display useful information alongside the current work area, would be preferable.

ELLA is a powerful design tool. It can be used to describe a design at a very abstract level which can then be refined until a gate level implementation is arrived at. Other tools, such as Silvar-Lisco's Helix simulator, can be used in a similar way but, rather than using an HDL, Helix uses Pascal for its high level description which necessarily implies limitations.

However, unlike Helix, ELLA is not integrated with other design tools. While access to the EASE database is possible through the ELLANET programming interface, there is no clear route to silicon from ELLA. Nor are there any libraries of cells or devices readily available although Plessey are developing a cell-based I.C. design tool called GATEMAP, which uses ELLA as a front end[8]. Clearly, if ELLA is to be used for teaching the steps of VLSI design, or indeed for commercial work, effort will be needed in developing libraries and interfaces.

At present ELLA is limited to textual input and output. There is a trend towards the graphical capture of designs, but it is possible that this will go out of fashion as flowcharts have in the area of software engineering. Furthermore, the envisaged facility to graphically display signals generated by a simulation would be a definite advantage. This is especially true at the more detailed levels of design.

5 CONCLUSIONS

Although there seems to exist no clear route to silicon from descriptions in ELLA, its use simply as an architectural design tool has hopefully been conveyed. With the current research effort focussed on bridging this gap, the advent of extremely powerful VLSI design tools cannot be far away.

REFERENCES

[1]   *The ELLA User Manual.* Praxis Systems plc, 20 Manvers St, Bath
[2]   Haynes, L. S. et al. (1982, 'A survey of highly parallel computing', *Computer* (Jan, 1982)
[3]   Jesshope, C. R., Rushton, A. J., Cruz, A., Stewart, J. M. (1987), 'The structure and application of RPA—a highly parallel adaptive architecture', *Highly Parallel Computers*, pp. 81–95. Elsevier
[4]   Jesshope, C. R., O'Gorman, R. and Stewart, J. M. (1988), 'A microprocessor array', to be published *IEE Proc Part E.*

[5]   Barton, I., 'Using an HDL at the analogue level', *Electronic Product Design* (Oct, 1987)
[6]   Stewart, J. M., *A Microprogramming and Simulation Environment for a Parallel Computer System*, M.Phil Southampton University (1988)
[7]   Jesshope, C. R., Miller, P. R., Yantchev, J. 'Programming with active data', *Proc. Parcella 88, Berlin* (1988)
[8]   Salmon, J. V. et al., 'Syntactic translation and logic synthesis in Gatemap', *Proc. IEE Colloquium on VLSI Systems Design* (Oct, 1987)

ABSTRACTS–ENGLISH, FRENCH, GERMAN, SPANISH

Using ELLA as a design tool

An overview of the ELLA system is given, followed by a description of how it has been used as an architectural tool to develop packet-switching networks for SIMD processor arrays. Finally, the merits and drawbacks of ELLA are discussed, with a summary on current trends in automated design.

Utilisation de ELLA comme outil de conception

Une vue générale du système ELLA est donnée dans cet article, suivie par une description de son utilisation comme outil d'architecture dans le développement de réseaux de commutation par paquets pour des réseaux de processeurs SIMD. En conclusion, les mérites et les inconvénients de ELLA sont discutés, avec un résumé des tendances actuelles dans la conception automatisée.

Nutzung von ELLA als ein Entwurfstool

Ein Überblick über ELLA wird gegeben, an die sich eine Beschreibung anschließt, wie es als Architektur-Tool genutzt wird, um Packet-Switching Networks für SIMD-Prozessor-Arrays zu entwickeln. Abschließend werden die Vor- und Nachteile von ELLA sowie in einer Zusammenfassung die gegenwärtigen Trends im automatisierten Entwurf diskutiert.

Utilización de ELLA como herramienta de diseño

Se ofrece un repaso del Sistema ELLA, seguido por una descripción de cómo se ha usado una herramienta arquitectónica para SIMD processor arrays. Finalmente se discuten los pros y los contras de ELLA, con un resumen de la tendencia actual del diseño automatico.

FUNCTIONAL MODELLING USING HELIX

E. G. CHESTER
Department of Electrical and Electronic Engineering, The University of Newcastle upon Tyne, England

INTRODUCTION

HELIX is a general purpose behavioural modelling package which has been made available to the academic community through the ECAD initiative. It allows the user to write models for devices which are not covered by vendors' libraries or for new designs, then to simulate these in conjunction with library components.

The device models are written in HHDL (hierarchical hardware description language) which is based on the Pascal programming language[1]. There are two major extensions to Pascal. Firstly, it is possible to run concurrent sub-processes, i.e. to have several 'procedures' executing in parallel. Secondly, net types are available in addition to normal variables. A net can have a value scheduled to appear on it at some future time, whereas variable assignments are instantaneous. Other features include the provision of special procedures for simple Boolean logic and register operations and for the parsing of input files.

The devices which are modelled in HHDL can be drawn as symbols in the SDS/CASS schematic entry package[2]. These symbols can then be inter-connected on a circuit diagram using a graphical editor. The circuit connect-ivity data can then be generated in the design database.

The circuit connectivity data combined with the HHDL model data is then used to produce Pascal source code for the simulator. This code, once compiled and linked with the HELIX libraries, can then be run to simulate the circuit. Circuit input signal values can be supplied interactively or in batch during the simulation, and the output signal values can be displayed in textual form or dumped to a file for later inspection as waveforms using the LOGAN post-processor[3].

Many users do not write their own models, but simply design circuits using commercial model libraries supplied by vendors. Although this allows complex circuits to be simulated faster than a full logic simulation, it does not exploit the full power of the HELIX package. The aim of this paper is to demonstrate the way in which HHDL can be used to create models, firstly for a commercial part for which (we assume) no model is available, and secondly for a new design which is being created 'top down' and has not yet reached the stage where the full structure is known.

```
MODULE UART;

USE REG PACK;
TIMEUNITS 8;  (* 10 ns *)

NETTYPE
    BOOLNET = BOOLEAN;
    DATAREG = REGISTER [7..0];
    DATABUS = 0..255;
    BITSTATE= (IDLE,START,B0,B1,B2,B3,B4,B5,B6,B7,STOP1,STOP2);
    CLKSTATE= 0..15;
(******************************************************************)
(***)   COMPTYPE IM6402;                                      (***)
(******************************************************************)
    INWARD TBR: DATABUS;
            SBS,EPE,PI,RRI,RRD,SFD,DRR,RRC,MR,CRL,CLS1,CLS2,TRC,TBRL: BOOLNET;
    OUTWARD TRO,PE,FE,OE,DR,TBRE,TRE: BOOLNET;
            RBR: DATABUS;
    INTERNAL TBREG,TREG,RREG,RBREG: DATAREG;
            RSTATE,TSTATE: BITSTATE;
            TCSTAT,RCSTAT: CLKSTATE;

SUBPROCESS
    (*--------------------------------------------------------------*)
    MRESET: UPON MR CHECK MR DO
            BEGIN
            ASSIGN 0 TO TCSTAT;    ASSIGN IDLE TO TSTATE;
            ASSIGN TRUE TO TRE;    ASSIGN TRUE TO TBRE;
            ASSIGN TRUE TO TRO;
            END;
    (*--------------------------------------------------------------*)
    TRLOAD: UPON TBRL CHECK TBRL DO
            BEGIN
            ASSIGN REGFROMINT(TBR,8) TO TBREG DELAY 9;
            ASSIGN FALSE TO TBRE DELAY 2;
            END;
    (*--------------------------------------------------------------*)
    TXCLK:  UPON TRC CHECK TRC DO
            IF TCSTAT=15 THEN ASSIGN 0        TO TCSTAT
                         ELSE ASSIGN TCSTAT+1 TO TCSTAT;
    (*--------------------------------------------------------------*)
    TXMIT:  UPON TRC CHECK TRC  DO
            BEGIN
            CASE TSTATE OF

        IDLE: BEGIN
                IF NOT TBRE THEN
                    BEGIN
                    ASSIGN TBREG TO TREG;        ASSIGN FALSE TO TRE;
                    ASSIGN TRUE TO TBRE;         ASSIGN START TO TSTATE;
                    ASSIGN 0 TO TCSTAT DELAY 1;  ASSIGN FALSE TO TRO;
                    END;
                END;

        START,B0,B1,B2,B3,B4,B5,B6:
                BEGIN
                IF TCSTAT=15 THEN
                    BEGIN
                    ASSIGN REGRBIT(TREG,0) TO TRO; ASSIGN SUCC(TSTATE) TO TSTATE;
                    ASSIGN REGRSHIFT(TREG,1,FALSE) TO TREG;
                    END;
                END;

          B7: BEGIN
                IF TCSTAT=15 THEN
                    BEGIN
                    ASSIGN TRUE TO TRO; ASSIGN STOP1 TO TSTATE;
                    END;
                END;

        STOP1: BEGIN
                IF TCSTAT=15 THEN ASSIGN IDLE TO TSTATE ELSE
                IF TCSTAT=7  THEN ASSIGN TRUE TO TRE;
                END;

            END (*CASE*)
            END (*TXMIT*);
    (*--------------------------------------------------------------*)
BEGIN
    ASSIGN TRUE TO TRE;    ASSIGN TRUE TO TBRE;    ASSIGN TRUE TO TRO;
END;
```

FIG. 1   Listing of the model file UART.HDL.

MODELLING A COMMERCIAL PART

The example chosen here is the IM6402 UART[4]. The model (shown in Fig. 1) is restricted to a subset of the capability of this device in the interests of clarity. Only the transmitter side of the device is modelled, with a fixed serial data format of 8 bits, no parity and one stop bit. Extension of the model to include further features should be straightforward; the required pins and registers are included in the version described here, though they are not used.

The first step is to decide on the data structure for the internal state information of the device. Some of this may be obtained from the block diagram on the data sheet, a simplified version of which is shown in Fig. 2, other details must be inferred. Since the transmitter is double-buffered, two registers TBREG and TREG are provided. These are of type REGISTER since it is easier to do bit operations on this type. The controller operation is not quite so obvious, but investigation of the timing diagram and description of operation reveals that the serial data is synchronously clocked at 1/16 of the rate of the TRC clock signal. A counter in the model must be used to divide by 16, hence a state variable called TCSTAT is used to store the current count from 0 to 15. This counter is reset at the beginning of each transmission. Every time this counter reaches 15, the controller transmits the next bit of serial data. It keeps track of which bit is currently being sent by using a further state variable, TSTATE. This is an enumerated type which can take on values which are descriptive of the state, e.g. START, B0, STOP1 etc.

Once the state information has been defined, the subprocesses which define the sequence of operations may be written. Initially, the device must be reset, hence the MRESET subprocess initialises the values of the state variables and outputs when the MR signal goes to logic 1. The TXCLK subprocess provides the divided-down clock signal by causing TCSTAT to increase on each rising edge of TRC.

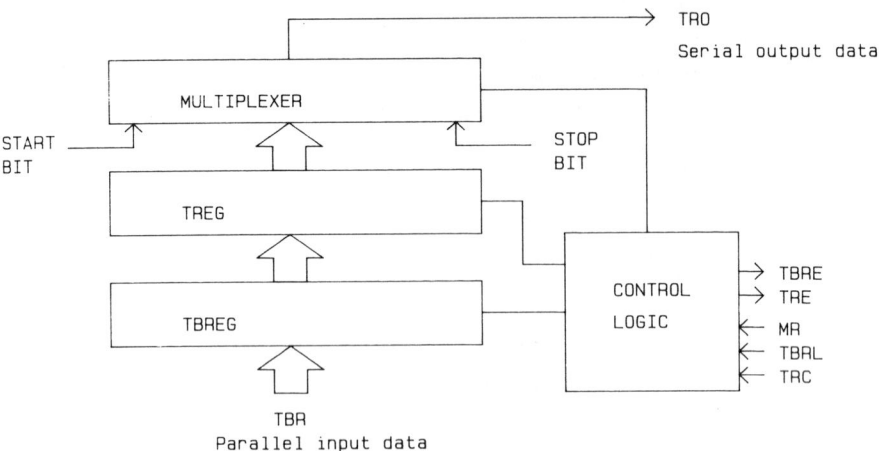

FIG. 2   Block diagram of UART transmit section.

Transmission of data is initiated by a rising edge on TBRL. This causes the TBRE signal to go low, indicating that TBREG contains data. The data is a subrange of integer rather than a bus of wires, since this is simpler to enter and display during simulation. Conversion to REGISTER type is performed by the function REGFROMINT. The transfer to TBREG takes place asynchronously, hence the TRLOAD subprocess is used to carry out this function. The TBRE signal, in addition to providing an output, is used to signal to the TXMIT subprocess that data is available for transmission.

The data sheet for the device indicates that output changes occur on the rising edge of TRC. The TXMIT subprocess is activated on every rising edge, therefore and a CASE statement is used to select the correct action according to the current state. If a state transition is required, a new value is assigned to TSTATE. If the current state is IDLE, a transition on the TBRE signal indicates that data is present in TBREG. This is loaded into TREG, the divide-by-16 counter is reset, and the START state is entered. The start bit and data bits B0 to B7 are then transmitted in sequence at intervals of 16 clock steps (transitions are conditional upon TCSTAT = 15) followed by a stop bit. On each data bit, the function REGRBIT extracts the required bit from TREG, then the shift operation is carried out by function REGRSHIFT. Note that the TRE signal goes high halfway through the transmission of the stop bit, hence this is conditional on TCSTAT = 7. The controller then returns to the IDLE state.

Some initialisation is carried out by the BEGIN..END block at the end when the simulator is first started. This is not strictly necessary, since the MRESET signal should achieve this, but it can be useful to aid debugging.

A circuit symbol for the UART was defined using CASS as shown in Fig. 3. This includes a pin corresponding to each signal input/output in the HHDL model. The simulation was run in TESTSYM mode (since a single component rather than a circuit is to be used) and signal values for TBR, TBRL, MR and TRC were controlled interactively.

The results of running the simulation are shown in Figs. 4 and 5. The input data word was 85 (binary 01010101). The transmit clock signal TRC was set up

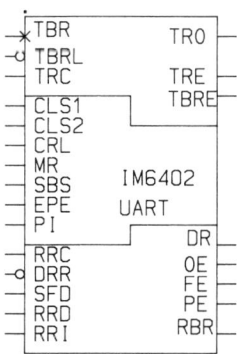

FIG. 3   CASS symbol for the UART.

using the REPEAT command to produce a square wave with a period of 100ns. It can be seen that serial data appears on the TRO output signal on the next rising edge of TRC after TBRL goes high, as expected, and that the serial data is correctly produced, with the appropriate status signals TRE and TBRE.

MODELLING A NEW DESIGN

This example shows how HELIX may be used to evaluate a simple digital filter design. The basic blocks are modelled at a fairly abstract level, but some realism is introduced by using integer values to represent signals.

The filter shown in Fig. 6 has an impulse response which is given by the following:

$$h_n = e^{-a(n-1)} . \sin \omega(n-1) \text{ for } n \geq 0$$

where the tap weight coefficients are:

$$A1 = 2e^{-a} . \cos \omega \text{ and } A2 = -e^{-2a}$$

The filter will therefore simulate the behaviour of a linear damped second-order system. A unit impulse at the input should produce an exponentially decaying sinusoid at the output.

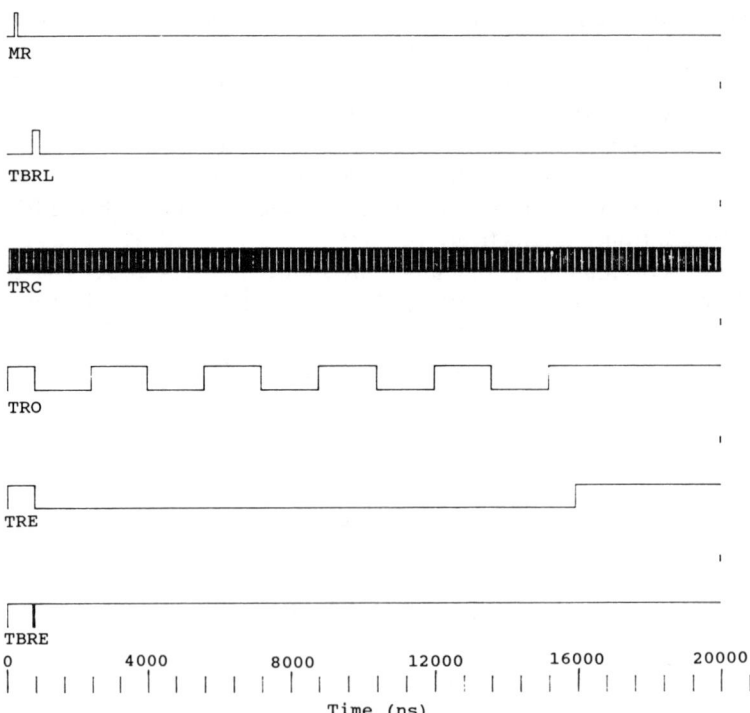

FIG. 4   UART waveform display from LOGAN.

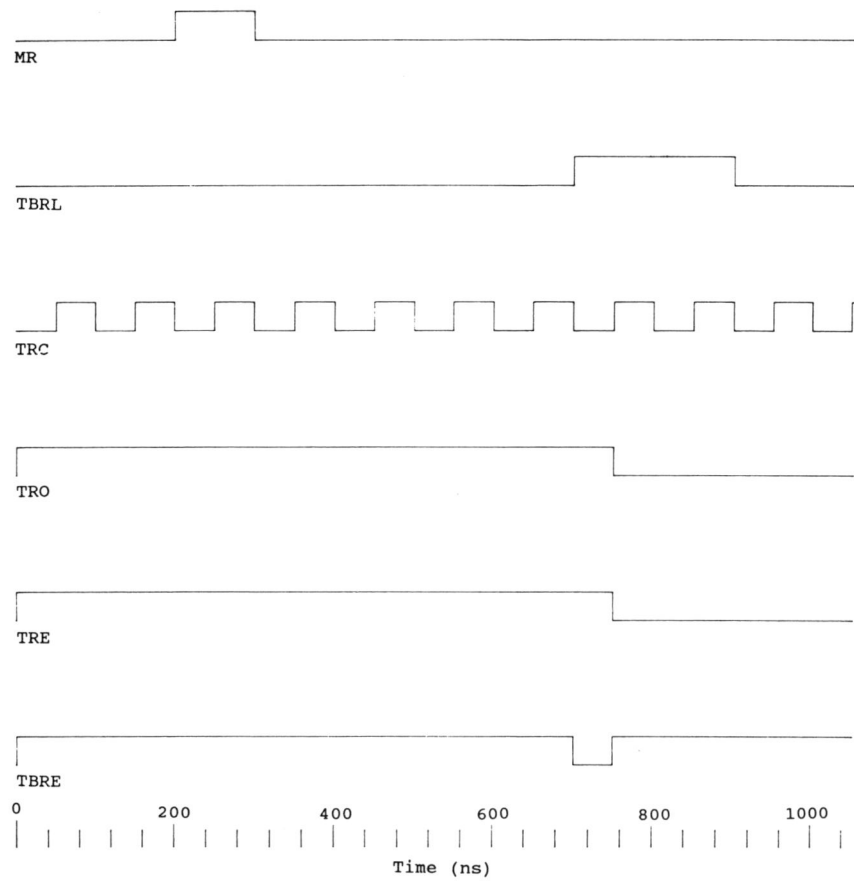

MR

TBRL

TRC

TRO

TRE

TBRE

Time (ns)

FIG. 5   Expanded view of UART waveform display.

The input signal will be generated by a 'signal generator' model rather than by changing signal values by using commands while running the simulator. The filter is connected to the signal generator as shown in the circuit in Fig. 7.

After drawing the symbols for the basic building blocks using CASS in symbol mode, the filter circuits can be constructed as shown. The models corresponding to the symbols must then be written in HHDL before simulation can proceed.

The HHDL models for the filter are contained in a file as shown in Fig. 8. Signals can be delayed by one clock period simply by using the DELAYUNIT block which assigns its input value to its output when a clock signal arrives. Addition is performed by ADDUNIT using the arithmetic addition operator. This happens asynchronously. Multiplication is carried out similarly by MULTUNIT, except that there is only one signal input; the other input

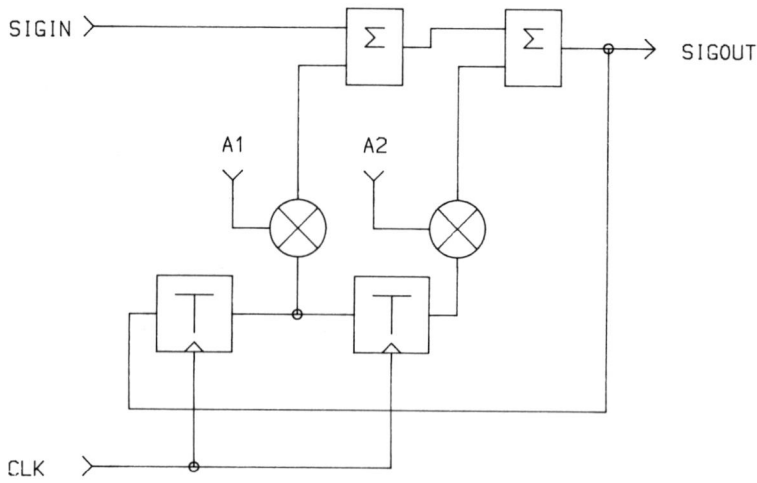

FIG. 6   CASS schematic of the digital filter.

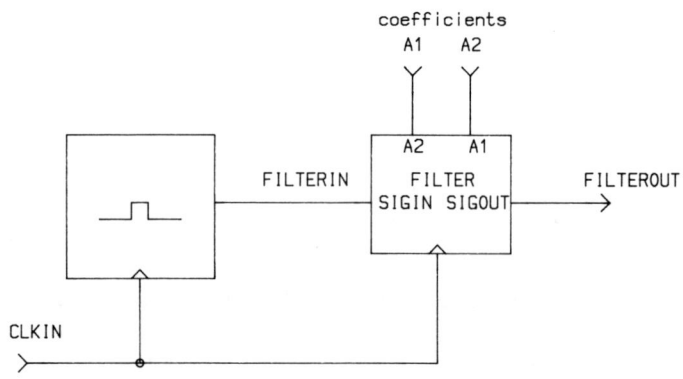

FIG. 7   CASS schematic of filter test environment.

is a coefficient which is normally kept constant throughout a simulation. Coefficients are given as real numbers to avoid confusion between signal and coefficient rounding errors. The input signal is a unit impulse provided by the SIGNALGEN block every 500 clock pulses.

The simulation result from the filter is shown in Fig. 9. The clock used is a square wave with a period of 20 time units (one time unit = 10^{-8} s). The output is a sinusoid as expected. The values of $A1$ and $A2$ are 1.97 and -0.98 respectively, corresponding to a decay constant $a = 0.01$ and natural frequency $\omega = 0.1$. The period of the sinusoid is thus $2\pi/0.1 = 62.8$ clock periods and the simulation result shows this to be so.

```
MODULE DIGFILTER;

TIMEUNITS 8; (*10 NANOSECONDS*)

NETTYPE
   SIGNALNET = -128..127;
   BOOLNET   = BOOLEAN;
   COEFFNET  = REAL;
(*******************************************************************)
(***)    COMPTYPE DELAYUNIT;                                  (***)
(*******************************************************************)
    INWARD   SIGIN:SIGNALNET;
             CLK  :BOOLNET;
    OUTWARD  SIGOUT:SIGNALNET;

    SUBPROCESS
    CLOCKING: UPON CLK CHECK CLK DO ASSIGN SIGIN TO SIGOUT DELAY 1;
BEGIN
ASSIGN 0 TO SIGOUT;
END;

(*******************************************************************)
(***)    COMPTYPE MULTUNIT;                                   (***)
(*******************************************************************)
    INWARD   SIGIN:SIGNALNET;
             COEFF:COEFFNET;
    OUTWARD  SIGOUT:SIGNALNET;

    SUBPROCESS
    MULTIPLY: UPON TRUE CHECK SIGIN DO ASSIGN ROUND(COEFF*SIGIN) TO SIGOUT;
BEGIN
ASSIGN 0 TO SIGOUT;
END;

(*******************************************************************)
(***)    COMPTYPE ADDUNIT;                                    (***)
(*******************************************************************)
    INWARD   SIGIN1,
             SIGIN2:SIGNALNET;
    OUTWARD  SIGOUT:SIGNALNET;

    SUBPROCESS
    ADD: UPON TRUE CHECK SIGIN1,SIGIN2 DO ASSIGN (SIGIN1+SIGIN2) TO SIGOUT;
BEGIN
ASSIGN 0 TO SIGOUT;
END;

(*******************************************************************)
(***)    COMPTYPE SIGNALGEN;                                  (***)
(*******************************************************************)
    INWARD   CLK:BOOLNET;
    OUTWARD  SIGOUT:SIGNALNET;
    VAR      COUNT:INTEGER;

    SUBPROCESS
    CLOCKING: UPON NOT CLK CHECK CLK DO
                 BEGIN
                 COUNT:=(COUNT+1) MOD 500;
                 IF COUNT=10 THEN ASSIGN 10 TO SIGOUT ELSE ASSIGN 0 TO SIGOUT;
                 END;
BEGIN
ASSIGN 0 TO SIGOUT;
COUNT:=0;
END;
```

FIG. 8   Listing of DIGFILTER.HDL.

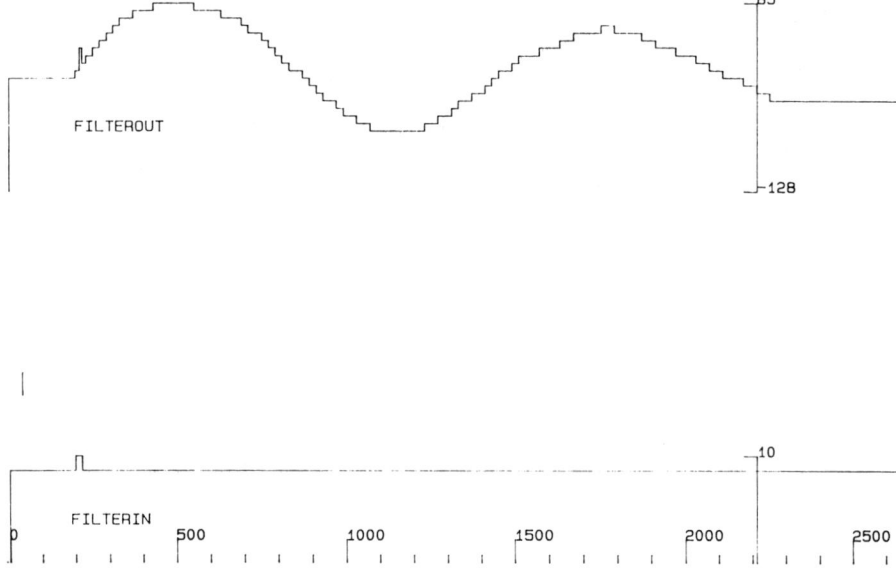

FIG. 9   LOGAN waveform display of digital filter output.

CONCLUSIONS

The HELIX language is a very flexible tool which allows circuits to be modelled as compositions of blocks, each of which may be described at a different level of abstraction. If an existing component is to be modelled, it can be described in as much detail as required. Unlike a simple gate level simulator, this allows simulation without total flattening of the hierarchy so that simulation time is very fast. If a component model is not provided by a vendor, it can be written by the user and included with the vendor's models (e.g. TTL 7400 series) to produce a composite circuit.

If a new design is to be modelled, the initial specification of the device can be captured by writing a high-level description of its function using HELIX, e.g. the adder circuit in the example. After evaluation of the design at this level, it may be decomposed top-down to produce a circuit, e.g. the adder could be constructed using single-bit adder blocks. The simulation could then be repeated using both the high level and decomposed versions to verify compliance with the specification. This decomposition process can be repeated until all the blocks are models of available hardware such as logic gates, flip-flops etc.

One drawback of the system is that any change to the schematic diagram or HHDL model requires a lengthy regeneration of the simulator code. It is wise, therefore, to control the variation of any parameters by signals which can be altered during the simulation run, if frequent changes are likely to be necessary.

REFERENCES

[1]   *SL2000/HELIX Command Reference Manual*, **2**, *Document No. HLX–2.2–004–1*,
      Silvar-Lisco Ltd., Menlo Park, California (November 1986)
[2]   *SL2000/Structured Design System Command Reference Manual*, **1**, *Document No.
      SDS–6.0–003–1*, Silvar-Lisco Ltd., Menlo Park, California (June 1986)
[3]   *SL2000/Simulation Base Command Reference Manual, Document No. SMB–1.0–003–1*,
      Silvar-Lisco Ltd., Menlo Park, California (September 1986)
[4]   'Universal asynchronous receiver and transmitter 6402 and 6402–1', *RS Data Library sheet
      no. 4046*, RS Components Ltd., Corby, Northants (March 1985)

ABSTRACTS–ENGLISH, FRENCH, GERMAN, SPANISH

Functional modelling using HELIX
A brief description of the Silvar-Lisco HELIX simulator is presented, followed by two example models. The first is a model of a standard component for which a data sheet is available. The second is a high-level model of a proposed circuit design for which no detailed information yet exists.

Modélisation fonctionnelle utilisant HELIX
Une brève description du simulateur Silvar-Lisco est présentée, suivie par deux exemples de modèles. Le premier est celui d'un composant standard pour lequel une feuille de spécifications est disponible, le second est un modèle de haut niveau pour un circuit proposé pour lequel aucune information détaillée n'existe.

Funktionsmodellierung unter Nutzung von HELIX
Eine kurze Beschreibung des Silvar-Lisco HELIX-Simulators wird vorgestellt, woran sich zwei Beispielmodelle anschließen. Das erste ist ein Modell für eine Standardkomponente, für die ein Datenblatt verfügbar ist. Das zweite ist ein High-Level-Modell eines vorgeschlagenen Schaltkreisentwurfs, für den bis jetzt keine detaillierte Information vorliegt.

Modelación funcional utilizando HELIX
Se presenta una breve descripción del simulador HELIX de Silvar Lisco, seguida de dos ejemplos, El primero es un modelo de un componente standard del que se dispone de una tabla de datos, El segundo es un modelo de alto nivel de un diseño de circuitos propuesto, del que no se dispone de información detallada todavia.

COMPUTER AIDED DESIGN OF SYSTEMS AT FUNCTIONAL LEVEL

N. D. DEANS, S. J. CHALMERS, C. PATERSON and A. JUTAGIR
Robert Gordon's Institute of Technology, Aberdeen, Scotland

INTRODUCTION

The development of a circuit or system from initial concept to final implementation can be viewed in terms of its 'design cycle'. The design cycle illustrated in Fig. 1 is typical, consisting of:

(i) The SPECIFICATION stage when the technical specification is devised.

(ii) The ARCHITECTURAL stage when the major subsystems are identified.

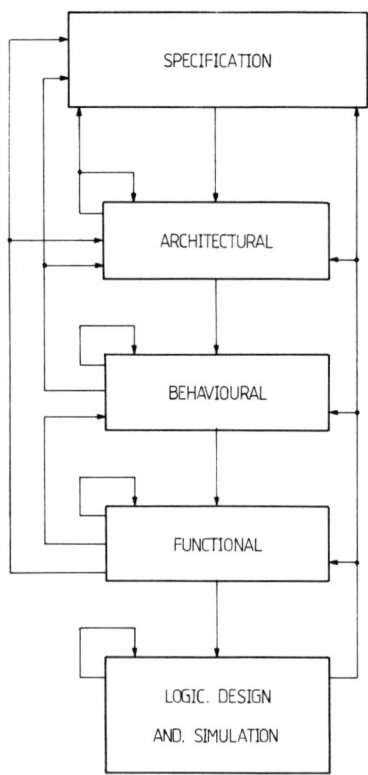

FIG. 1   The design cycle.

(iii) The BEHAVIOURAL stage at which the operations of the major sub-systems are examined.

(iv) The FUNCTIONAL design stage when subsystems are developed using circuit blocks.

 (v) The LOGIC design stage at which detailed circuit design takes place.

Existing CAD tools are largely directed at the logic design stage, and well-known software packages including inter alia SPICE, HILO and HELIX are used as computer-aided verification tools to examine the behaviour of circuits already designed by engineers. Circuit faults can be identified and corrected— provided that the design decisions made at the earlier stages in the design cycle are not flawed. To minimise the number of iterations within and around the design cycle, it is imperative that the validity of the design at the early stages is proved.

The hardware and software facilities installed in departments of electronic and electrical engineering under the DTI ECAD initiative considerably extended the range of tools available to academic staff to assist in the formation of good undergraduate electronic circuit and system designers.

This paper describes two high-level behavioural modelling facilities developed to run within the Silvar-Lisco SL2000 suite of programs. The facilities enable designers to simulate systems at block level and confirm their behaviour *before* detailed circuit designs are attempted. They are:

(a) An algorithmic state machine validation tool.

(b) A digital signal processing system design tool.

EDUCATIONAL AIMS

The objectives in developing the state machine and digital signal processing verification facilities were five-fold, viz:

  (i) to encourage a top-down design approach

 (ii) to encourage thinking at an abstract level

 (ii) to allow student engineers to examine the behaviour of complex systems

(iv) to encourage student-centred learning

 (v) to ease the understanding of difficult systems.

ALGORITHMIC STATE MACHINE VERIFICATION

Algorithmic descriptive techniques are conveniently used by systems designers to define the behaviour of digital state machines. Algorithmic State Machine (ASM) charts, first documented by Clare[1] and more recently detailed in teaching texts[2] provide a symbolic representation of the state transitions and output functions of a state machine. The ASM chart of the controlled modulo-3 counter shown in Fig. 2(a) illustrates the use of the three basic descriptive symbols, viz., the State Box, the Decision Box and the Conditional Output Box. The counter remains in state A whilst the input HOLD is true; when HOLD becomes false, the circuit cycles around states A, B, C, A, B ...., changing state on receipt of an external clock signal. An output signal HHELD is true asserted if the circuit is being held in state A and an output signal IHC is true asserted during the time the circuit is in state C.

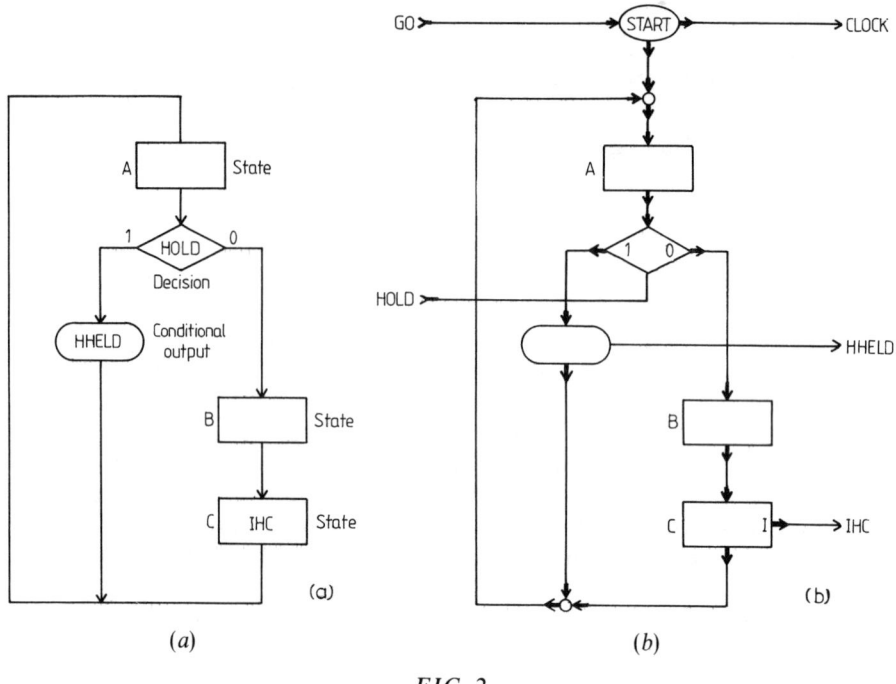

FIG. 2

The behaviour of ASM has been studied using the simulation facilities provided by the Silvar-Lisco HELIX suite of CAD programs. A library of descriptions of the primitive elements in ASM charts has been created. Symbolic representations of the elements have been generated using the Computer-Aided Schematic System (CASS) package and each symbol is supported by a behavioural description written in the Hierarchical Hardware Description Language (HHDL). The CASS library symbols developed closely resemble the symbols originally described by Clare. However, the requirement for input and output signals to be explicitly defined within a HELIX simulation, implies that corresponding input and output pins must exist in the CASS representation. Input qualifiers and output functions appear therefore as additional nets. The CASS ASM chart corresponding to the conventional ASM chart of Fig. 2(a) is shown in Fig. 2(b).

In addition to the three basic symbols, two additional library primitives have been developed. A Start Box is used to initiate a sequence of operations and a Connection Box is used to interlink nets.

The HHDL behavioural descriptions are compiled using SIMCHK into a .HDB library database. An ASM chart entered using CASS is submitted to the Net List Extractor (NLE) program for checking and the compilation of the net list. The expansion program HIDEX is used to reduce any elements expressed in a hierarchical fashion to their primitive representations. SIMLINK is invoked to merge the structural and behavioural descriptions of the elements and their connectivity into a simulator for that design. The results of the simulation run can be examined using LOGAN, the logic analysis and presentation program.

The facility provides for large (> 100 states) state machines, machines with varying clock rates and systems involving several serial-linked or parallel linked machines to be investigated.

DIGITAL SIGNAL PROCESSING SYSTEMS

There is a distinct lack of design tools available to implement systems-level design for signal processing. Several programs such as SIG[3] and MATLAB[4] are available. Signals are operated upon by a series of commands, followed by parameters. Graphical data entry is not provided. Work at The Polytechnic of Central London Centre for Microelectronic Systems Applications has demonstrated that HELIX can be used to model digital signal processing functions. The work described here used HELIX and HHDL to develop models to allow digital filter design and verification.

Digital signal processing systems require remarkably few basic building blocks. Those necessary are the unit delay, adder and multiplier. All could be simply implemented as behavioural models using HHDL. Unfortunately HELIX has input and output arrangements designed specifically for digital circuit analysis. Input is via successive DEPOSIT and REPEAT commands, and output is via the LOGAN logic analyser program. Neither of these is suitable for the current application. Alternative input and output methods have been developed.

INPUT

The input method adopted was via a program SIGNAL which simulates a signal generator. Parameters such as waveform, amplitude, frequency and sampling frequency are user-selectable. When these have been selected, a sampled-data sequence is generated and stored in a sequential file. The user can store up to ten signal generators for subsequent inclusion in simultations.

OUTPUT

A program DISPLAY was written to allow simultaneous plots of up to four waveforms. The waveforms are displayed as four colours on an overlay plot. This format has particular advantages in digital filter design, since direct comparisons of magnitude and phase can be made. At present this program produces only time domain plots. Data is read from an output file or a signal generator file. The format of all data files is therefore standardised.

The interface between these programs and the HELIX simulator is very simple. The signal generator is represented by a model which, upon initialisation, reads data from the sequential file and stores this in an array. The display model consists of code which writes data to a sequential file. Both models are

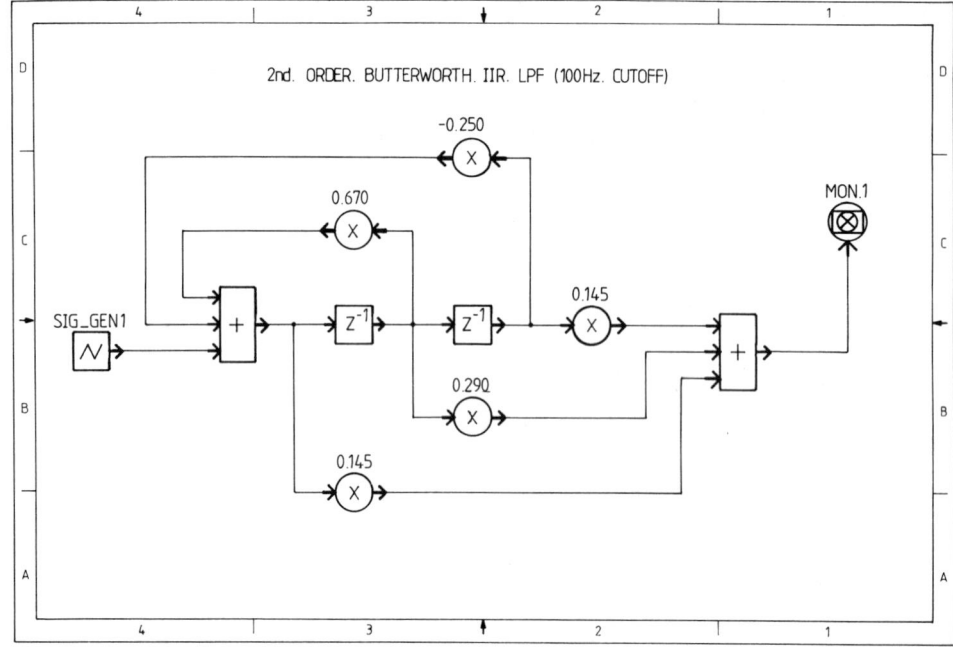

FIG. 3   Digital filter implementation.

activated by an internal simulation clock. In symbolic form these models are represented by a signal generator block and a monitor point. Using CASS, these, and the other signal processing symbols can be connected to form signal processing functions. A typical application is illustrated in the signal flow graph structure of Fig. 3. In addition, the hierarchical nature of HELIX can be used here to create useful building blocks, which in turn can be incorporated into increasingly complex systems. An example might be cascadable first and second order IIR filters or n-tap sections of an FIR filter.

The steps required to run a signal processing simulation can be summarised are as follows:

(i)   Enter the system signal flow graph via CASS. Run NLE and HIDEX to verify correct connections.

(ii) Set up input signals by running signal generator program (SIGNAL.BIN).

(iii) Run SIMLINK to execute simulation.

(iv) Execute DISPLAY.BIN to view simulation results.

Successive simulations can be run by repeating steps (ii)–(iv).

CONCLUSIONS

There is clearly a demonstrable need for the provision of CAD tools to enable electronic circuit and system designers to validate and evaluate systems described at block level. The development of tools for investigating the behaviour of algorithmic state machines and digital signal processing systems has been described. The schemes make use of the structural and behavioural modelling facilities offered within the Silvar-Lisco SL2000 suite of programs. The work was carried out using Apollo DN3000 workstations.

Details of the models can be supplied by authors Chalmers and Deans, School of Electronic and Electrical Engineering, Robert Gordon's Institute of Technology, Schoolhill, Aberdeen.

ACKNOWLEDGEMENTS
The authors wish to acknowledge the assistance given by Mrs. J. Cox and Mr. W. Rawles in preparing this paper, and Professor G. Cain, Mr. M. Griffiths and Mr. Paul Hughes of PCL for discussions on signal processing with HELIX.

REFERENCES
[1]   Clare, C., *Designing Logic Systems Using State Machines*, McGraw-Hill (1972)

[2]   Green, D., *Modern Logic Design*, Addison-Wesley (1986)

[3]   Lager, D. L., Azevedo, S. G., 'SIG-A General-Purpose Signal Processing Program' *Proc IEEE* **75** No. 9, pp. 1322–1332 (Sept, 1987)

[4]   Moler, C. et al., *Matlab Users Guide*, The Math Works Inc, Sherborn, MA, USA (1987)

ABSTRACTS–ENGLISH, FRENCH, GERMAN, SPANISH

Computer aided design of systems at functional level
Existing CAD tools are largely used to examine and verify the behaviour of circuits already designed by engineers. This paper describes high-level behavioural modelling facilities developed to simulate algorithmic state machines and digital signal processing systems and so confirm their behaviour before detailed circuit design work is undertaken.

Conception assistée par ordinateur de systèmes au niveau fonctionnel
Les outils existant da CAO sont largement utilisés pour examiner et vérifier le comportement de circuits déjà conçus par les ingénieurs. Cet article décrit les facilités de modélisation du comportement de haut niveau développées pour simuler des machines d'état algorithmique et des systèmes de traitement de signaux numériques et d'ainsi confirmer leur comportement avant d'entreprendre un travail de conception détaillée des circuits.

Computergestützter Entwurf von Systemen auf dem Funktional-Niveau
Existierende CAD-Tools werden weithin genutzt, um das Verhalten von Schaltungen zu prüfen und zu verifizieren, die durch Ingenieure entworfen wurden. Dieser Artikel beschreibt High-Level-Verhaltensmodellierungs-Möglichkeiten, die entwickelt wurden, um algorithmische Zustandsautomaten und digitale Signalprozessor-System zu simulieren und so ihr Verhalten zu bestätigen, bevor die detaillierte Schaltungsentwurfsarbeit durchgeführt wird.

Sistemas de diseño asistido por ordenador a nivel funcional
Las herramientas CAD existentes se utilizan mucho para examinar y comprobar la conducta de circuitos diseñados por Ingenieros. Este articulo describe la modelacion de conductos de alto nivel desarrollados para simular los algoritmos de estados de máquina y sistemas de procesamiento de señales digitales y también confirma su conducta antes de que el circuito detallado se haya realizado.

A GRAPHIC DISPLAY TOOL FOR THE HILO LOGIC SIMULATOR

TREVOR P. HOPKINS and DAVID MATHIESON
Computer Science Department, University of Manchester, England

The Department of Computer Science at the University of Manchester has a long tradition of teaching in the computer hardware engineering area. Consequently, the Department has been using a number of the facilities provided through the ECAD Initiative.

In particular, the HILO*[1,2] logic simulator has been used for second year undergraduate teaching. The ability of HILO to simulate the operation of a complete microcomputer system based on the Motorola 68000 microprocessor[3] was highly attractive, as the M68000 family was used as the basis for many exercises, both hardware and software.

This choice was particularly appropriate as the Department has recently acquired a large number of *Sun* workstations to support first and second year laboratory work. At that time, relatively few of the software packages supported under the ECAD Initiative were available on Suns; however, HILO was found to be satisfactory for our purposes.

Unfortunately, the normal output from HILO consists of text files containing signal names and values in character form. No graphical display tool was available which worked in our environment. This paper describes an interactive graphical display system capable of showing the results of a simulation run on a Sun workstation screen.

INTRODUCTION TO HILO

HILO is a logic simultation system supplied by GenRad Ltd, which is supported on a wide range of machines and operating systems. It represents the system to be simulated in terms of logic gates, modelling the timing delays found in real gates. The input format is a textual description, using the *Gate Description Language*. A variety of technologies (TTL, NMOS, CMOS, ECL) can be modelled. There are some additional features which represent the additional delays caused by the capacitance of the interconnecting wires. HILO can also model the 'transfer' gates available in some technologies.

As the gate-level description of a large system may be very complex, HILO also provides a *Functional Modelling Language* (FML). This is particularly suitable for LSI and VLSI devices, such as microprocessors. FML descriptions are given in terms of a (relatively) conventional programming language. A

---

*HILO is a Trademark of GenRad Corporation.

library of standard devices is supplied, containing both gate-level and functional descriptions.

In order to excite the circuit under test, 'input' waveforms can be described using the *Waveform Description Language*. This can also specify the expected responses from the simulated circuits, so that HILO can be used to verify the operation of a circuit. HILO also supports the generation of test vectors, including automatic test pattern generation for sequential circuits.

REQUIREMENTS FOR GRAPHICAL OUTPUT
Two forms of output from the HILO simulator are provided. The *Display* format gives a text representation of selected signals at certain times (for example, whenever any signal changes). The display output representation is fixed before each simulation run, including which signals are to be monitored. This means that if the states of further signals need to be known for some reason, another simulation run is required. This may be very time-consuming. Further, as the display output is intended to be printed on a line printer, only a limited number of signals can be viewed at a time.

In an attempt to overcome these problems, an alternative output mechanism is provided. The simulated values on some (or, more typically, all) of the wires in a circuit may be *captured* during a simulation run. A separate program called DISPRO* is used to select which signals are to be printed, and at which (simulated) times. DISPRO also provides facilities to detect specified *trigger* conditions, so that signals may be printed only in the region of a trigger condition. Unfortunately, DISPRO still uses a textual output format, representing values as streams of characters.

USER INTERFACE DESIGN
The primary requirement of the HILO display tool was the ability to show the simulated waveforms captured from a HILO run on the workstation screen in a graphical manner. Displayed output using the conventional 'timing diagram' form was regarded as the most important. This form seems to be more readily understandable for many purposes than the character-based 'state' display. However, the 'state' display form is also provided. The ability to select the signals to be displayed, and the (simulated) time scales to be used was considered very important.

In designing the user interface, the 'logic analyser' metaphor was used, so that the interface would be familiar to computer hardware engineers. For example, extensive use of 'buttons' was made for controlling the display parameters. Nevertheless, the availability of a windowing environment on the workstations led to the adoption of techniques not applicable to conventional instruments; these include pop-up menus and multiple overlapping windows. For example, these techniques are used to select the signals to be displayed.

---

*DISPRO is a Trademark of GenRad Corporation.

Support for triggering conditions was included, modelled on those available in DISPRO.

Figure 1 shows a view of a workstation screen with the HILO display tool in operation. The main sub-window of the 'logic analyser' display shows a simulation of a MC68000 microprocessor executing the first few instructions after a system reset. The smaller sub-window to the right allows the user to

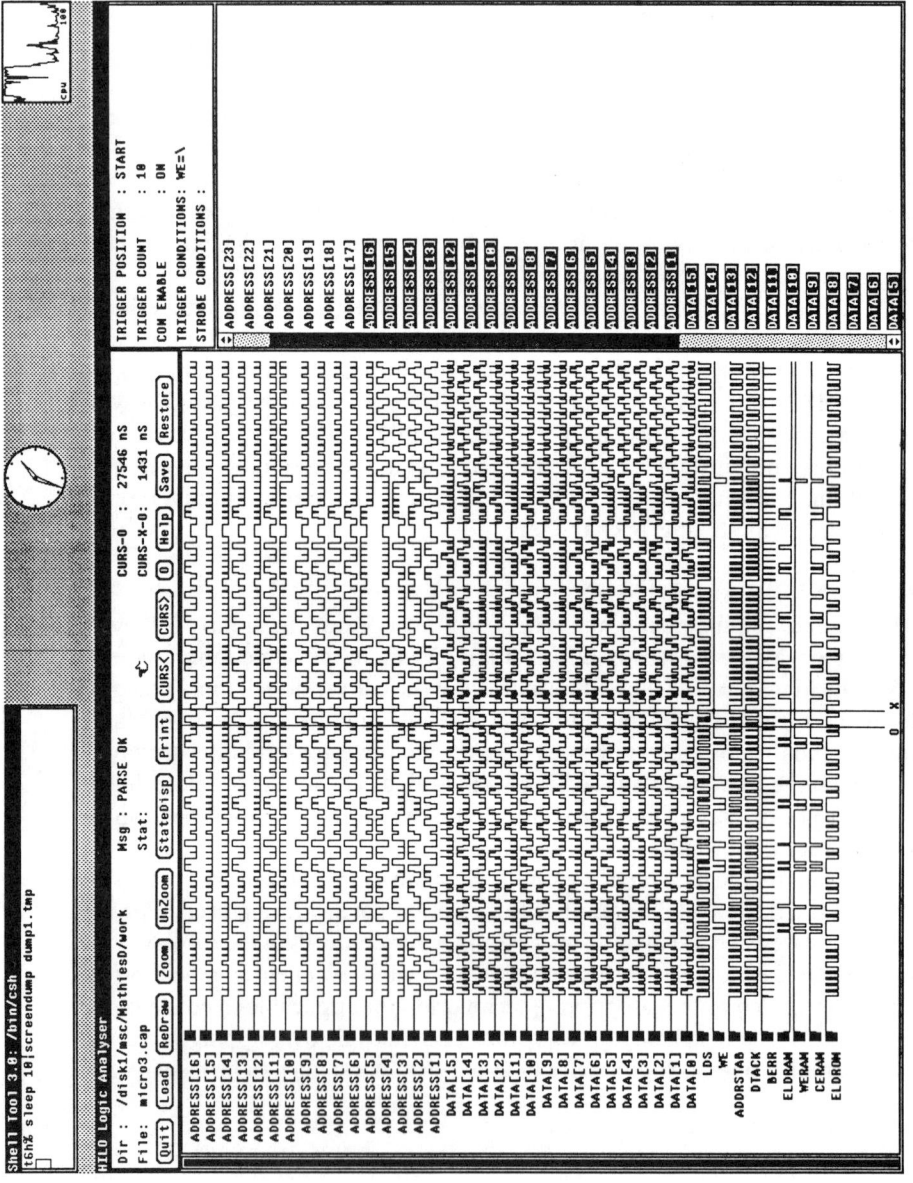

FIG. 1   HILO display tool in operation.

select which signal names are to be displayed. Both these two larger windows can be scrolled, if there are too many signals to fit into the available space on the screen. A small sub-window at the top right allows the user to select the desired trigger conditions. Finally, a sub-window at the top of the 'logic analyser' display contains a number of 'push-buttons' to select various functions.

Graphical output in **Postscript***[4,5] form was also a requirement. This format is widely used by laser printers, and is the preferred way to get hard-copy output in the Department. Figure 2 illustrates the output available on the laser printer for the same simulation illustrated in Fig. 1.

IMPLEMENTATION

The HILO graphical display system is implemented on top of the Sun Micro-systems operating system (**SunOS**†[6]), which is based on UNIX‡[7]. This made it possible to use the lex lexical analyser generator and the yacc parser generator tools, which are available under the UNIX operating system, for parsing the captured data from HILO. The rest of the system is implemented using the C programming language[8]; this allows convenient interfaces to the **SunView**[9] system, which supports interactive applications using the **SunTools** window manager.

Further details of the implementation of the logic display system can be found in Ref. [10].

CONCLUSION AND FUTURE WORK

The HILO display tool described here works well with the sort of simulations used in the undergraduate teaching laboratories in the Department.

A small number of limitations have been noticed in use. The number of signal names is fixed; this is a compile-time constant in the program. At present, only 2000 signal names can be used. The order in which signal names appear in the display is determined by the order in which these names appear in the HILO simulation, and cannot be altered. These restrictions could be removed by using more sophisticated data structures within the program. Finally, all files used by the display tool must be in the same directory, and fixed file names are used. This is occasionally an irritation.

Other display modes analogous to those found in logic analysers could also be added. The use of a 'state map' display might be useful in some applications.

In view of the manufacturer-specific nature of the **SunView** system, it would be sensible to consider re-implementing the graphics routines using an operating system independent window manager, such as **NeWS***[11] or **X Windows**†[12]. Unfortunately, manpower for this task has not been available.

***Postscript** is a Trademark of Adobe Systems Inc.

†**SunOS**, **SunView** and **SunTools** are Trademarks of Sun Microsystems Inc.

‡UNIX is a Registered Trademark of AT&T Bell Laboratories.

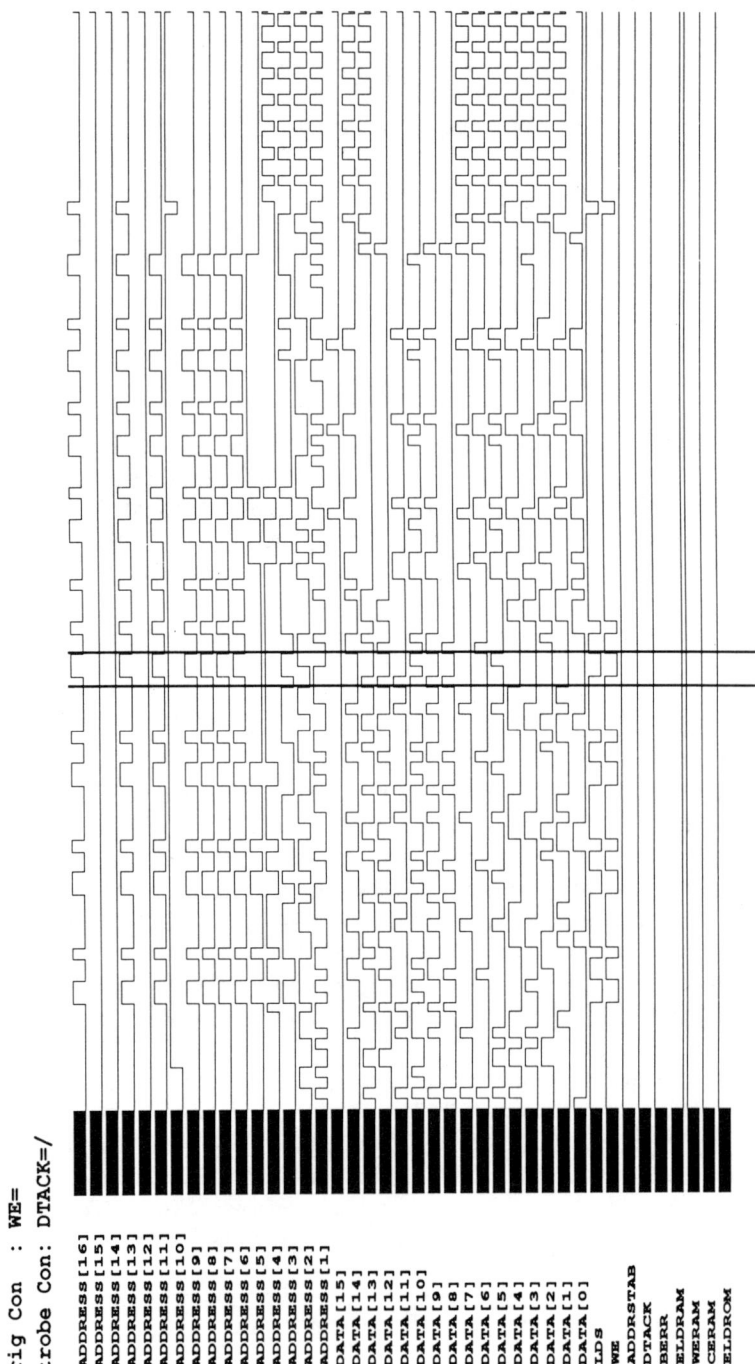

FIG. 2 Postscript output from HILO display tool.

REFERENCES

[1] *HILO Reference Manual.* GenRad Inc (1984)
[2] Blundell, B. G., Daskalakis, C. N., Heyes, N. and Hopkins, T P. *An Introductory Guide to Silvar Lisco and HILO Simulators,* Macmillan Press (1987)
[3] Kane, Gerry, *68000 Microprocessor Handbook,* Osborne/McGraw-Hill (1981)
[4] *Postscript Language Reference Manual,* Adobe Systems Inc., Addison-Wesley (1985)
[5] *Postcript Language Tutorial and Cookbook,* Adobe Systems Inc., Addison-Wesley (1985)
[6] *Sun Workstations Manuals,* Sun Microsystems Inc. (1986)
[7] Bourne, S. R., *The UNIX System,* Addison-Wesley (1982)
[8] Kernigham, B. W. and Richie, D. M., *The C Programming Language,* Prentice-Hall Inc. (1979)
[9] *SunView Programmers Guide,* Sun Microsystems Inc. (1986)
[10] Mathieson, D., *A Software Logic Analyser for the HILO Logic Simulator,* Master's thesis, Department of Computer Science, University of Manchester (1987)
[11] *Sample NeWS Manual,* Sun Microsystems Inc. (1987)
[12] Scheifler, R. W. and Gettys, J., 'The X Window System', *Transactions on Graphics,* ACM (1986).

---

*NeWS is a Trademark of Sun Microsystems, Inc.
†X Windows is a Trademark of Massachusetts Institute of Technology.

ABSTRACTS–ENGLISH, FRENCH, GERMAN, SPANISH

A graphical display tool for the HILO logic simulator
One of the ECAD programs supported by the UK 'ECAD Initiative' is the HILO logic simulator. This paper describes the design and implementation of a graphical output tool for HILO. This tool presents the results of a simulation run on a Sun workstation screen in the manner of a logic analyser, and can also convert the display format to Postscript for output to a laser printer.

Un outil d'affichage graphique pour le simulateur logique HILO
Un des programmes supportés par l'initiative ECAD en Grande-Bretagne est le simulateur logique HILO. Cet article décrit la conception et la réalisation d'un outil graphique d'affichage pour HILO. Cet outil présente les résultats d'une simulation sur l'écran d'une station de travail SUN à la manière d'un analyseur logique et peut aussi convertir le format d'affichage en Postscript pour sortie sur imprimante laser.

Ein Grafic-Display-Tool für den HILO-Logik-Simulator
Eines der ECAD-Programme, die durch die UK-ECAD-Initiative unterstützt werden, ist der HILO-Logik-Simulator. Dieser Beitrag beschreibt den Entwurf und die Implementation eines Grafik-Ausgabetools für HILO. Dieses Tool präsentiert die Resultate eines Simultationslaufes auf dem Schirm einer Sun-Workstation in der Art eines Logikanalysators und kann auch das Display-Format zum Postcript für den Ausgang zu einem Laser-Drucker konvertieren.

Una herramienta gráfica para el simulador lógico IIlLO
Uno de los programas ECAD soportados por 'UK ECAD Initiative' es el simulador lógico HILO. Esta articulo describe la implementación y el diseño de una herramienta de salida gráfica para HILO. Esta herramienta presenta los resultados de une simulación corriendo de forma de analizador lógico en una pantalla de une estación de trabajo SUN, y puede también convertir el formato de salida de pantalla a una salida de impresora laser.

USE OF EXISTING CELL LIBRARY AND SOFTWARE TOOLS IN A SILICON COMPILATION ENVIRONMENT

C. C. JONG, R. A. BERGAMASCHI, M. ZWOLINSKI, D. J. ALLERTON and K. G. NICHOLS
Department of Electronics and Computer Science, University of Southampton, England

1 INTRODUCTION

Advances in VLSI technology allow highly complex systems to be fabricated on a chip. Silicon compilers, which can translate a high-level description of a circuit into layout, provide an efficient way for generating VLSI systems of large complexity. They enable system designers to concentrate their efforts on system-level and architecture-level designs without detailed knowledge of IC design and processing technology. This paper reports on the current status of the SCHOLAR Silicon Compiler[1-5] developed at Southampton, with emphasis on the use of existing cell libraries and software tools in conjunction with the compiler. SCHOLAR is a general silicon compiler for the synthesis of concurrent VLSI systems from behavioural descriptions. No specific target architecture is assumed and any systems that can be mapped into the general model of a data path and a control unit can be generated by the compiler. The behavioural description is first translated into a register-transfer form called Intermediate Code (ICODE), which constitutes the input to the functional simulator and to the synthesis processes. The output of the synthesis processes is a technology-independent representation of the design in terms of parameterised modules and Boolean equations. The modules and Boolean equations are then implemented by using standard cells and a netlist of cells is obtained. By using a standard cell library and a placement-and-routing program, the physical layout is generated.

Figure 1 shows the compilation environment, where the rectangular boxes show the application software tools and the elliptical circles show the design at different representation levels. The software tools enclosed in the dash-line box are obtained under the ECAD initiative, which include the SMS3 Cell Library of Silicon Microsystems Ltd., the SL2000 suite of Silvar-Lisco Inc. and the HILO3 simulator of GenRad Ltd.

2 SCHOLAR HARDWARE DESCRIPTION LANGUAGE

The SCHOLAR Silicon Compiler translates a behavioural description of an integrated system into layout. The behaviour of a system is described in the SCHOLAR Hardware Description Language (HDL)[1,2] which is a structured

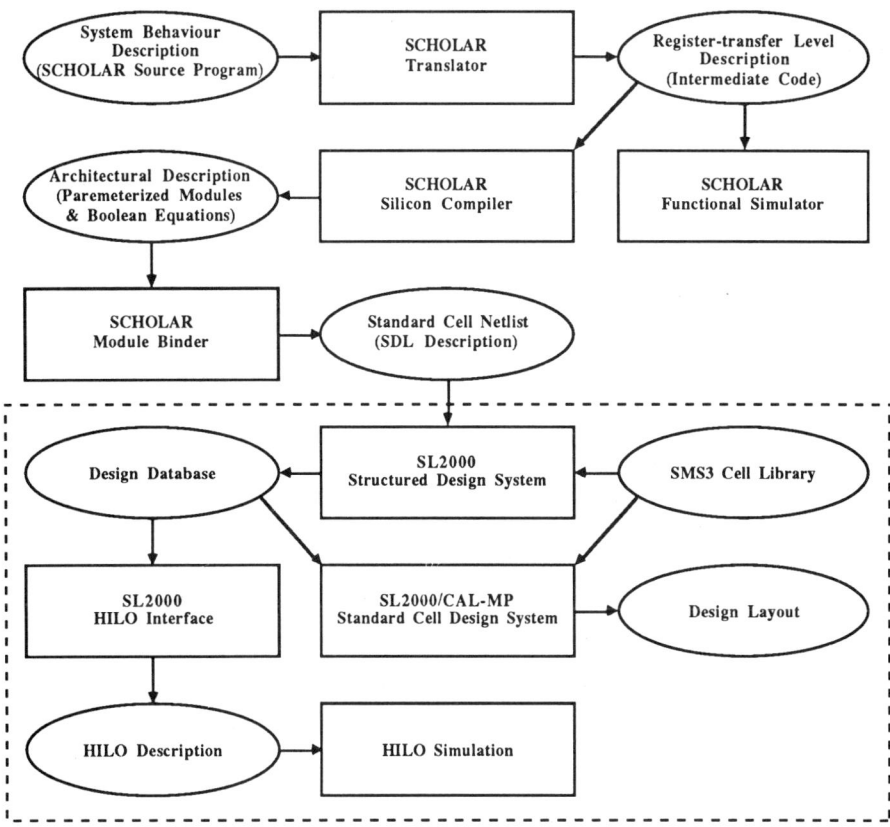

FIG. 1   SCHOLAR silicon compilation environment.

high-level programming language and has the ability to specify the behaviour of a system independent of its implementation. The language allows the specification of sequential as well as parallel processes and also allows modular designs by providing procedural declarations. The following is a section of a SCHOLAR description.

```
WHILE (a > 0) AND (x < y) DO
$(                              // starting a sequential section
    x := y + z
    d := e XOR f
    $[                          // starting a parallel section
        IF (d AND e) OR f
        THEN a := a − b
        ELSE a := a − c
        f := a > 10
    $]                          // ending the parallel section
$)                              // ending the sequential section
```

Both sequential and parallel processes can be specified in SCHOLAR HDL. Parallelism is either described explicitly or is extracted by the compiler during the optimizing stage. Arbitrary nesting of sequential and parallel declarations is allowed.

A design can be partitioned into several units, which can be declared as MODULES or MACROS in SCHOLAR HDL. The main difference between a module and a macro in SCHOLAR is that multiple calls to a module correspond to multiplexed access to the module but multiple calls to a macro cause the macro to be replicated physically in the design.

Variables in SCHOLAR have no data type associated with them. They are declared as bit strings with a specified width and hence, can be mapped directly into registers of desired width. Access to individual bits or part of a register is possible through aliasing of variables.

The description of system behaviour is translated into a register-transfer form at the first stage of the compilation process and the described system can be simulated functionally at the register-transfer level. The SCHOLAR Functional Simulator[3] provides a method for verifying the functional correctness of the description. Errors in translating a system design from its specification into the behavioural description can be detected by the functional simulation.

3 SYSTEM SYNTHESIS

From the register-transfer description of the system, an Internal Graph Representation (IGR) is derived. The directed graph is optimised so that the described system is operated in the minimum number of clock cycles. The parallelism overlooked by the designer is extracted by the compiler during this optimisation process. The optimised graph contains the information for synthesising the control and the data path of the system. The synthesis procedure consists of the following steps:

(i) *Allocation of storage elements to the variables declared in the description* A storage element is allocated to each variable. The feature of an element depends on the set of operations involving that variable. For example, if a variable is initialised with a constant, then the register for this variable will contain the necessary set/reset feature.

(ii) *Allocation of functional units to operators* Each operator in the description is implemented as an abstract functional unit in the form of a parameterised module. For example, an add operation results in an adder of the size equal to the maximum width of its operands. In order to save the silicon area, the following optimizations are performed during the allocation of functional units. First, if a variable X is involved only in the operation $X := X + 1$, then X is allocated to an incrementer (counter) instead of a register plus an adder. Second, operations in SCHOLAR may share operators. The operations which use similar operators and are never executed at the same time are allocated to a shared operator. Extra multiplexors are inserted to the inputs of the shared operator and extra control signals are needed to control the multiplexors. Hence, operators are shared only if silicon area can be saved, i.e. only if the area becomes smaller after the sharing.

(iii) *Definition of necessary interconnections* Interconnections are assigned point to point. Multiplexors are added at the inputs of the functional units and the registers as necessary.

(iv) *Control synthesis* The controller is constructed from the final IGR. The nodes of the IGR correspond to the states of the controller and the arcs to the state transitions. Hence, the controller consists of a sequencer plus a set of Boolean equations describing the control signals. The controller is able to control sequential processes as well as parallel processes.

4 MODULE BINDING

The output of the synthesis process is a netlist of technology-independent parameterised modules describing the data path plus a set of minimised Boolean equations describing the control signals. After the netlist and the control signals have been obtained, the use of existing standard cell libraries and powerful placement-and-routing software tools provides an efficient way to the physical implementation of silicon designs. Moreover, the use of predefined cells allows the estimation of the chip's area and delay at an early stage, which is used to guide the synthesis steps. For example, if a SCHOLAR statement cannot be executed in the required time, then it is partitioned and executed in more clock cycles depending on the estimated delay.

The netlist and the control signals are mapped into standard cells by the module binding process, which consists of the following two steps:

(i) *Construction of the parameterised modules using single-bit cells* Each module consists of the following parameters: type (function), size (number of bits), input/output signals and other attributes depending on the type of the module. For example, a 4-bit counting-up counter may require set/reset function and tristate output. The module binding process is to construct the required counter according to these parameters.

The construction of parameterised modules consists of selecting the appropriate cells from the cell library and establishing the inter-cell connection. If a required cell is not available in the library, the Module Binder will construct the cell using available cells. Optimisation is also performed during the binding stage, particularly for those modules with constants as their inputs.

(ii) *Implementation of the Boolean equations using standard cells*[5] The set of minimised Boolean equations, which describes the control/signals, is implemented by the logic synthesis and technology mapping system embedded in SCHOLAR. The Boolean equations are first minimised into a sum-of-products form and then transformed into a multilevel representation (Boolean network) using factorisation techniques based on weak-division and kernel extraction.

The Boolean network is implemented using library cells by mapping each Boolean factor (or group of factors) into one or more cells selected from the library according to an area or speed optimisation criteria. The logic synthesis and technology mapping system can be used as a stand-alone module generation system or integrated in the SCHOLAR environment.

An important characteristic of the Module Binder is that it can be easily adapted to other cell libraries.

5 LAYOUT GENERATION

Once the parameterised modules and the Boolean equations have been implemented using standard cells, a netlist of cells is obtained and the layout can be generated by automatic placement-and-routing programs.

Currently, the standard cell library adopted by SCHOLAR is the SMS3 3μm CMOS cell library[6] of Silicon Microsystems Ltd. The Silvar-Lisco SL2000 suite is used to generate the layout after the netlist is completed by SCHOLAR. Both the SMS3 cell library and the SL2000 suite are obtained under the ECAD initiative. They are used in conjunction with SCHOLAR because they are appropriate and available. The use of available cell library and placement-and-routing programs in the SCHOLAR compilation environment provides an efficient way to the physical implementation of silicon design.

To generate the layout, the netlist of standard cells is described in the Structured Design Language (SDL)[7] of Silvar-Lisco. The SDL description is then loaded into the design database through the SL2000 Structured Design System (SDS)[8]. The layout is generated by using the standard cell design system CAL-MP[9] of the SL2000 suite, which performs the placement and the routing processes.

There is another advantage in using the SL2000 suite. Once the SDL description is loaded into the database, translations from the SDL description into other formats are possible. One of the translations which is particularly useful is the one into HILO description. Although SCHOLAR has its own functional simulator for verifying functionally the design correctness, the HILO simulator is used for gate-level simulation to perform more precisely timing analysis and for verifying the correctness of the silicon compilation process during the development stage. The GenRad's HILO3[10] simulator, which is also obtained under the ECAD initiative, is used currently.

6 CONCLUSIONS

The SCHOLAR Silicon Compiler can synthesise VLSI systems from behavioural descriptions. The parameterised modules and Boolean equations, generated by the Compiler, are mapped into standard cells. Placement-and-routing programs are used to generate the layouts. Using the SCHOLAR Silicon Compiler, the cost and time for designing VLSI systems can be reduced significantly. Through the ECAD initiative, the availability of the suitable cell library and the layout tools used in the SCHOLAR Compilation Environment allows the final stage of the SCHOLAR project to be completed earlier with small financial cost.

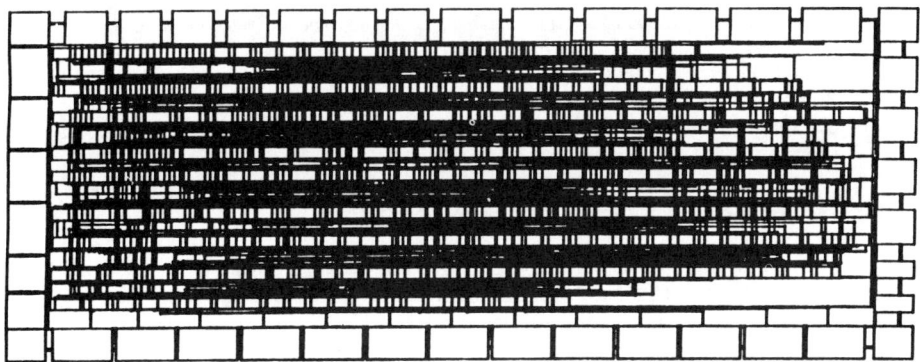

FIG. 2 Layout of the CRT controller circuit designed by using SCHOLAR. Dimension: x-direction 8.25 mm, y-direction 3.77 mm, area 31.1 mm².

The circuits compiled by SCHOLAR include, among others, a microprocessor, a graphics controller and an arithmetic processor. Results comparable with manual implementations were obtained in most cases. Figure 2 shows the layout of a CRT controller circuit compatible with the Motorola MC 6845 CRT Controller. The design consists of more than 800 cells or 3000 equivalent-gate (one equivalent-gate is equivalent to one 2-input NAND gate).

In general, the SCHOLAR project has successfully demonstrated that by putting together behavioural and structural synthesis algorithms with an efficient cell library and layout tools, it is possible to generate good quality VLSI design from high-level description in a short time, and efforts can be concentrated on the system design rather than the tedious and error-prone circuit and layout design. From an academic perspective, this allows students to concentrate on the architectural issues, and from an industry perspective, it allows new designs to be built quicker and cheaper, hence, to reach the market earlier and be more competitive. Additionally, SCHOLAR allows rapid implementation of ASICs and allows ASICs to be designed by system designers, who only have little knowledge of IC design.

7 REFERENCES

[1]   Allerton, D. J. and Currie, A. J., 'SCHOLAR: another approach to silicon compilation), *Pro ICCAD 84, Santa Clara* (November 1984)

[2]   Allerton, D. J., Batt, D. A. and Currie, A. J., 'A VLSI design language incorporating self-timed concurrent processes', *European Design Automation Conf.* (1984)

[3]   Allerton, D. J. et al., 'Functional simulation as an adjunct to silicon compilation', *IEE Int. Conf. on CAE* (December 1984)

[4]   Bergamaschi, R. A. and Allerton D. J., 'A graph-based silicon compiler for concurrent VLSI systems', *IEEE Compeuro 88 Conf., Brussels* (April 1988)

[5]   Bergamaschi, R. A., 'Automatic synthesis and technology mapping of combinational logic', *IEEE ICCAD-88, Santa Clara* (November 1988)

[6]   Silicon Microsystems Ltd., 'SMS3 CMOS Cell Library Users Handbook', August 1986

[7]   Silvar-Lisco Inc., *SDL—The Structured Design Language Reference Manual*, Document No. M–004-4 (July 1984)

[8]   Silvar-Lisco Inc., *SL2000/Structured Design System—Command Reference Manual*, Document No. SDS–6.0–003–1 (June 1986)
[9]   Silvar-Lisco Inc., *SL2000/CAL-MP 11.2—The Standard Cell Design System—Command Reference Manual*, Document No. CMP–11.2–003–1 (July 1986)
[10]   GenRad Ltd., *HILO-3 User Manual* (June 1986)

ABSTRACTS–ENGLISH, FRENCH, GERMAN, SPANISH

Use of existing cell library and software tools in a silicon compilation environment
This paper describes the SCHOLAR silicon compiler and the use of the Silvar-Lisco design suite, the HILO3 simulator and a cell library, obtained under the ECAD initiative, in a silicon compilation/module generation environment. SCHOLAR is a general silicon compiler for synthesis of concurrent VLSI systems from behavioural-level descriptions.

Utilisation de librairie de cellules existantes et outils logiciels dans un environment de compilateur de silicium
Cet article décrit le compilateur de silicium SCHOLAR et l'utilisation de l'ensemble de conception Silvar-Lisco, du simulateur HILO3 et d'une librairie de cellules, obtenus par l'initiative ECAD, dans un environment compilateur de silicium/générateur de modules. SCHOLAR est un compilateur de silicium général pour la synthèse de systèmes VLSI à partir de descriptions au niveau du comportement.

Nutzung einer existierenden Zellbibliothek und von Software-Tools im Rahmen eines Silicon-Compilers
Dieser Artikel beschreibt den SCHOLAR-Silicon-Compiler und die Nutzung der Silvar-Lisco-Entwurssoftware des HILO3 Simulators und einer Zellbibliothek, erhalten unter der ECAD-Initiative, in einem Compilation/Modul-Generationsrahmen. SCHOLAR ist ein allgemeiner Silicon Compiler für die Synthese von VLSI Systemen, ausgehend con Beschreibungen auf Verhaltensniveau.

Utilización de biblioteca de células y software existente en un entorno de un compilador de Silicio
Este articulo describe el compilador de Silicio SCHOLAR y el paquete de diseño de SILVAR-LISCO, el Simulador HILO3 y biblioteca de células obtenidas bajo la iniciativa ECAD en un entorno compilador de Silicio/generación de módulo. SCHOLAR es un compilador de Silicio general para sintesis de sistemas VLSI concurrentes a partir de descripciones del nivel de comportamiento.

NOTES

REAL-TIME IMAGE PROCESSING PROJECT

S. P. HARRISON, A. D. HOUGHTON and M. SEED
Department of Electronic and Electrical Engineering, University of Sheffield, England

An electronic circuit has been developed to enhance an existing real-time image processing system, called RAPAC[1]. The design and constructional work was carried out by a final year student during his project, which centred around the design of a custom gate-array chip. This chip has been fabricated, and successfully included into the system. Extensive and essential use was made of CAD facilities acquired as part of the 'ECAD Initiative' — specifically, custom integrated circuit design, and printed circuit board layout tools.

The image processing system is used to automatically monitor road traffic, for example at a complex road junction, where data is acquired using a black and white video camera. The system comprises a number of pipelined hardware preprocessing circuits (including this one) under software control, which perform such tasks as binary thresholding, background subtraction, noise reduction and feature extraction, in real-time. Higher level information concerning image features (cars, lorries, etc) is then passed to a host computer where tasks such as vehicle tracking, counting and classification are performed, again in real-time.

The new circuit removes noise in the image, each frame of which comprises a 256×256 array of pixels. The input to the circuit is a binary image, corrupted with noise in the form of small groups or single pixels. The frame rate is 50 Hz, and the circuit operation is equivalent to a rate of 5 million instructions per second on localised pixel data, outputting a substantially noise-reduced image of the same format.

The noise reduction process is carried out by four cascaded 'filter' circuits implemented on a single gate-array chip. Each filtering process can be thought of as passing a 3×3 square pixel window over every part of the image, updating the old pixel values as it goes. Thus the new value of the centre pixel under the 3×3 window is determined by its old value, and those of its eight nearest neighbours. If greater than n out of the nine pixels are set in a given window (where n is a number between 0–9 and is programmable for each filter) then the outputted centre pixel is set. Otherwise the centre pixel is reset. If $n < 3$, then features in the image will tend to be enlarged by the operation, whereas if $n > 3$ a thinning operation will result. By appropriate choice of n for the four filter stages, such that alternate thinning and then growing operations are used, the noise can be removed.

This project has demonstrated clearly the benefits of a hierarchical design approach for such circuits. This ranged from the circuit board level specifications in terms of functional and host system interface requirements, to the gate-array chip on the circuit board and the five levels of hierarchy within the chip itself (down to logic gate level). A 'top-down' design plus 'bottom-up' implementation strategy was used with careful logic simulation, and design for testability at each stage. The gate-array circuit used approximately 1500 logic gates (out of 2800 available on the chip) and was fabricated using 3 μm cmos technology, by Micro Circuit Engineering Ltd., Tewkesbury, Glos.

The experience gained in custom integrated circuit design has been very encouraging. In particular, the facility to fabricate designs moderately cheaply is very good for student and staff motivation alike.

REFERENCE

[1]   Elphinstone, A. C. et al., 'RAPAC — a high speed image processing system', *Proc. IEE*, **134**, Part E, No. 1, pp. 39–46 (Jan., 1987)

OPTIMAL DESIGN OF A NOVEL CLASS OF SWITCHED-CAPACITOR FILTERS USING SWAP

R. E. MASSARA and A. T. YOUNIS

Department of Electronic Systems Engineering, University of Essex, England

This study concerns the optimal design of a new switched-capacitor (SC) filter structure which features a number of practical advantages including reduced active and passive component count and hence reduced chip area, reduced on-chip capacitance spread, and low sensitivity to primary and parasitic components. The design technique makes novel use of the SWAP SC analysis utility which forms part of the Silvar-Lisco SL2000 suite of programs.

INTRODUCTION

The Bruton transformation is a widely used approach to the realization of low-sensitivity RC-active filters based on a complex impedance scaling of a prototype passive RLC ladder network. All prototype circuit impedances $Z(s)$ are transformed to $Z_T(s)$ with

$$Z_T(s) = \frac{K}{s} Z(s),$$

where K is a scaling constant chosen by the designer. This impedance transform is conveniently implemented by direct application to the elements in the prototype network and it is easily shown that a resistance R transforms into a capacitance $C = K/R$, whilst an inductance L transforms into a resistance $R = K.L$. This latter result, the elimination of inductors in favour of resistors, constitutes the essential rationale for the Bruton transform method. Prototype capacitors are slightly more problematical; applying the Bruton transform to a circuit capacitance C yields

$$Z_T(s) = \frac{K}{s^2 C} = \frac{1}{s^2 D},$$

where $D = C/K$ is the parameter value of what may be regarded as a new component produced by the transformation and referred to as a frequency-dependent negative resistance (FDNR). In practice, the FDNR elements are simulated by RC-active sub-networks using operational amplifiers. Massara et al.[1,2] showed that FDNR-RC active filters derived from lossy LCR-ladder networks can be designed by means of a non-uniform predistortion technique to exhibit low capacitance spread and low sensitivity to component parameter variations — both features important for integrated circuit implementation of the devices.

The integrated circuit realization of active filters has been significantly enhanced by the advent of the switched-capacitor (SC) concept and many techniques for the design of SC filters have been proposed. A common feature in many of these design techniques is the derivation of an SC filter from an analogue prototype by mapping into the discrete time domain. This approach has the significant advantage that the extensive literature available on passive filters, including comprehensive tabulations of filter designs, can be used directly thus eliminating the need to solve the synthesis problem, having determined a suitable approximation. The approach to be proposed here is based on the SC simulation of FDNR elements obtained by Bruton transformation.

The present study has led to the development of an SC filter design technique based on the SC realization of lossy-FDNR subcircuits. The approach produces designs which are optimal in terms of active and passive component count, which feature very low sensitivity, and which offer minimised capacitance spread — an important requirement for efficient integrated circuit realization.

DEVELOPMENT OF LOSSY-FDNR SC FILTER

In the approach adopted here, 'lossy' RLC lowpass prototype ladder filters (in which the shunt capacitors are deemed to have associated parallel conductances representing loss) are optimally designed for realization in SC form. Fig. 1 shows (a) a third-order example of the prototype LCR

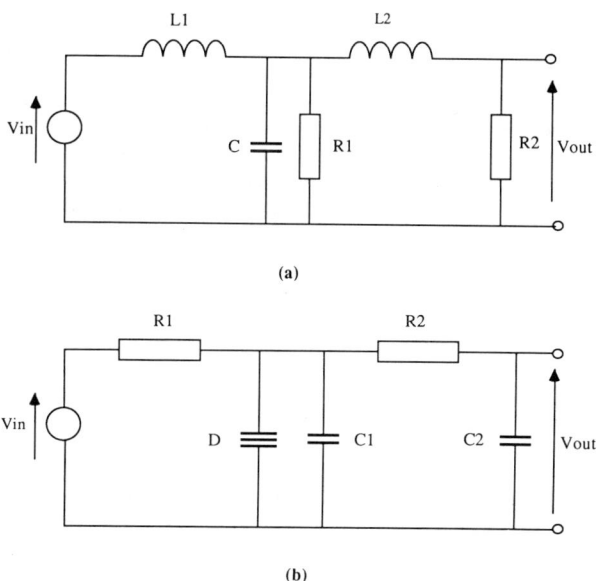

FIG. 1 *(a) RLC lossy prototype (b) Bruton transformation.*

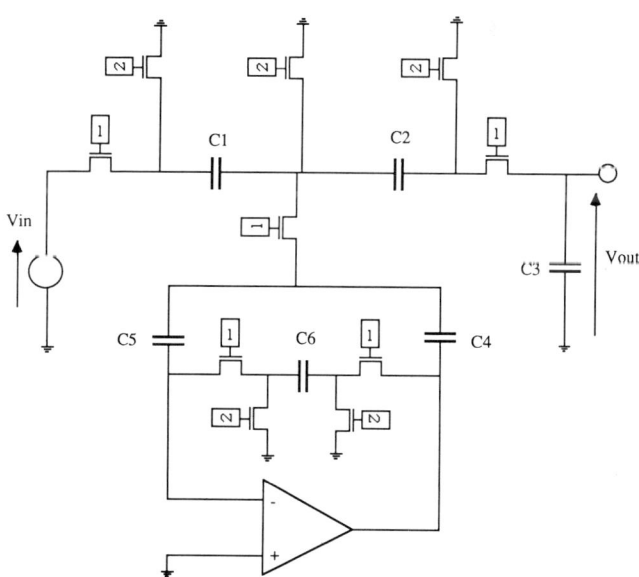

FIG. 2 *SC-FDNR 3rd order lowpass filter.*

network, and (b) its Bruton-transformed equivalent. One approach to realizing an FDNR-based circuit in SC form is to make use of one of the standard s-plane to z-plane frequency transformations. Initial consideration of the commonly-used bilinear and backward transformations showed that the bilinear-transformed filter is sensitive to parasitic capacitances whereas the backward transform yields a network which, in the case of the present structure, is relatively insensitive to parasitics. It was thus decided to use the backward transform. Fig. 2 shows the SC filter resulting from this transformation, when applied to the prototype structure of Fig. 1(b). For a given specification, the component values for the original and transformed networks are found by numerical methods based either on solving the set of non-linear equations derived by matching pre-warped specification coefficients with the explicit literal network coefficient expressions or by direct numerical optimisation.

DESIGN OF LOSSY-FDNR-BASED SC FILTERS BY OPTIMISATION

Design complexity increases with increasing order, and the prewarping technique becomes difficult for high-order filters. In addition, the problem of capacitance spread becomes critical — as in all SC filter design — as order increases. The use of a direct optimisation approach was introduced next as a means of dealing with the capacitance spread problem whilst avoiding the need to pre-warp specified transfer functions. The technique exploits one of the Silvar-Lisco SL2000 SC analysis utilities, SWAP, in an unusual way. It is worth noting that this design technique can be applied to any SCF structure. It has been found that the method satisfactorily deals with the problem of reducing on-chip capacitance spread and total circuit capacitance — of vital importance to final silicon area. Fig. 3(a) shows the passband characteristic produced when the filter is designed in this way to satisfy a standard third-order Chebyshev specification. Sensitivity data derived through SWAP, and plotted in Fig. 3(b), confirm that primary circuit sensitivities are low.

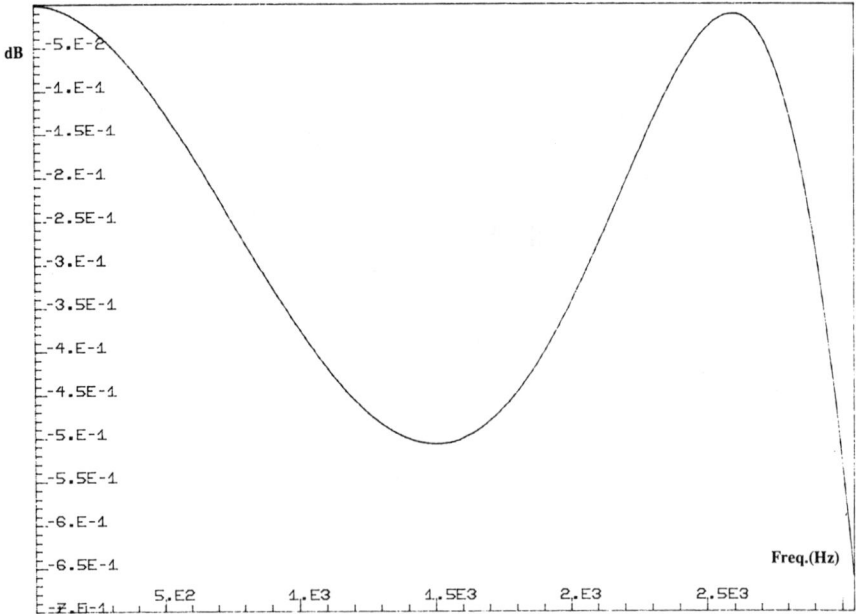

FIG. 3   (a) Optimised frequency response

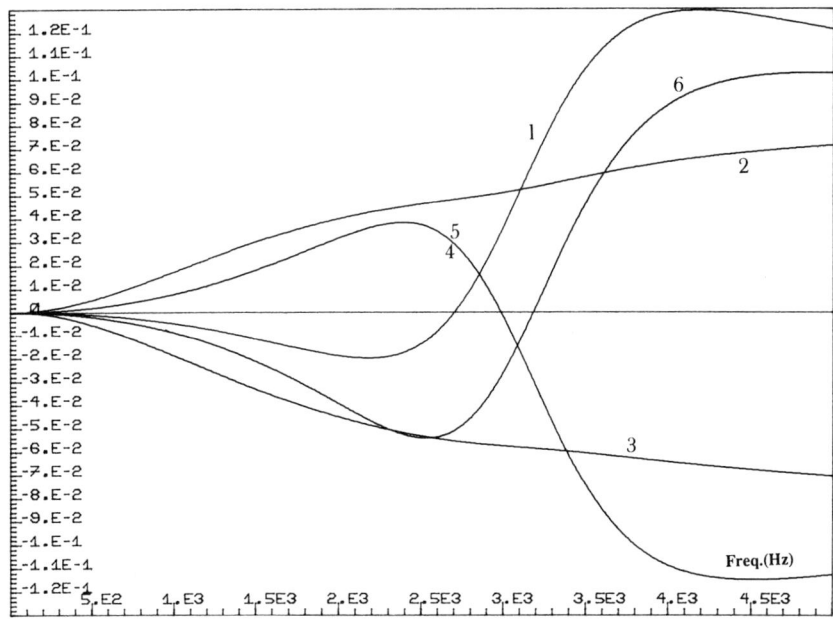

FIG. 3 (b) Sensitivity of frequency response to capacitors (C_i).

PARASITIC CAPACITANCE AND CAPACITANCE SPREAD

The effect of parasitic capacitances is crucial in the design of SC filters. In the new structure, bottom-plate parasitic capacitance insensitivity is featured, and only the comparatively small top-plate parasitic capacitances are significant. Simulated frequency responses and circuit sensitivity characteristics show that the top-plate parasitics have negligible effect on the frequency response and that the sensitivities associated with the parasitic capacitances are insignificant compared with the sensitivity of the primary circuit capacitors.

The price of the reduction of op. amp. count in the filter structure is an increase in capacitance spread and hence in total capacitance. A very attractive reduction of the capacitance spread is available by modifying the op. amp. feedback branch, which usually contains the smallest capacitance value. This modification, when appropriate, doubles the smallest capacitance value and thus decreases the capacitance spread by approximately one half. The modification does not invalidate the feature of bottom-plate parasitic insensitivity.

REFERENCES

[1] Massara, R. E. and Al-Najjar, A. R., 'FDNR realization of all-pole lowpass filters', *IEE Proceedings*, Pt. G., **128**, No. 4, pp. 195–197 (1981)

[2] Massara, R. E. and Al-Saqer, A. Y., 'On the optimal design of FDNR-derived bandpass active filters', *Proc. 7th European Conference on Circuit Theory and Design — ECCTD '85*, Prague, pp. 621–624 (Sept., 1985)

THE UNIVERSITY OF SURREY MULTI-PROJECT CHIP

BARRY M. COOK
Department of Electronic and Electrical Engineering, University of Surrey, Guildford, England

For effective teaching of full-custom IC design, it is desirable that student designs should be fabricated and tested. A potential difficulty is the associated cost, since ideally these costs should be comparable with a normal project budget. This difficulty can be overcome by the use of a multiproject chip for student project work, as pioneered by Mead and Conway[1] long ago. This note describes a six-project harness designed for use in connection with the ISIS software and the MCE (Microcircuit Engineering Ltd.) 3-micron process.

The scale of project feasible within a typical student project time uses between 400 and 1000 transistors and occupies around 1 mm² of silicon. With some allowance for test points within the circuit, it has been found necessary to provide between 10 and 20 I/O connections for a project.

A suitable overall chip size is 4.0 mm by 4.3 mm, mountable in a 48-pin package. A harness has been designed for this chip size that will accommodate six student projects, each having a design area of 1030 by 1350 microns. For larger projects (funds permitting) two or three of the project areas may be combined by deleting the intermediary guard bands giving areas of 2100 by 1350 microns and 3170 by 1350 microns respectively. The harness provides that the projects are isolated from each other and the common circuitry, so that even a major design error (e.g. a power-to-ground short circuit) in one project will not affect the operation of other projects. By applying appropriate signals to 'activation' pins, projects can be powered up and connected to the harness main bus.

The I/O drivers linking this bus to the outside world provide:

(i)   4 simple inputs for data including versions for TTL or CMOS switching thresholds with simple or Schmitt-trigger characteristics.

(ii) 4 special inputs each having 2 pads, one with CMOS threshold Schmitt-trigger characteristic and the other with a pullup resistor and TTL threshold Schmitt-trigger characteristics. These may be used as a CMOS switching threshold input, a TTL switching threshold input with pullup resistor, a reset input (using an external resistor and capacitor) or a clock generator using an external RC network or quartz crystal.

(iii) 12 tri-state Inputs/Outputs, the state and direction of which can be controlled by the project in 3 nibbles each of 4 bits. The outputs will drive CMOS, TTL (2 standard loads) or LEDs (to 20 mA).

In addition, each project has access to 2 dedicated pads to allow analogue input/output or experiments with I/O drivers to be accommodated.

The structure and cells described have been designed, combined with experimental projects, and fabricated via Micro-Circuit Engineering at Tewkesbury on a 3-micron single-poly, double-metal process. The chips worked first time, and behaved in accordance with the predictions of the ISIS simulator. The maximum clock speed through the harness is in excess of 30 MHz, which is adequate for undergraduate projects.

Yield for the harness was about 80%. The yield for any individual project will be less than this, but most chips will have some working projects. In our set of 20 test chips we found several working examples of each project. For a standard fabrication run we would expect 20 packaged chips in about 10 weeks. The multiproject chip has also been used in connection with under-graduate projects, by Dieter Gollmann at the University of Karlsruhe.

This harness is now distributed by the Surrey ECAD leadsite as part of the auxiliary software associated with the ISIS package.

A fuller account of this Multiproject chip will appear elsewhere[2] and further information can be obtained from the author.

REFERENCES
[1]   Mead, C. and Conway, L., *Introduction to VLSI Systems*, Addison-Wesley (1980)
[2]   Massara, R. E. (Ed.), *Design and Test Techniques for VLSI and WSI Circuits*, Peter
      Peregrinus, forthcoming. Chapter by Cook, B. M. and Forbes, R. G., 'ISIS in the
      educational environment'.

THE USE OF THE ECAD SOFTWARE FOR RESEARCH

T. I. PRITCHARD, D. TAYLOR, J. RACZKOWYCZ and P. HALLAM
Department of Electrical and Electronic Engineering, The Polytechnic, Huddersfield, England

The Higher Education ECAD initiative in 1985 resulted in universities and polytechnics having
access to a wide range of industry standard electronic computer aided design software. This has
resulted in a number of collaborative research projects being initiated in academic establishments
which have been made possible only by the availability of these software packages. This note
describes three typical postgraduate projects in the areas of VLSI and ECAD carried out at this
polytechnic using the ECAD facilities.

1   *Self test of semi custom devices (with Micro Circuit Engineering Ltd., Tewkesbury)*
This project is concerned with the design of novel semi-custom cell-backgrounds, which are
inherently two-pattern self-testing. This should enable the testing times and hence costs of ASICs to
be drastically reduced.

   The work is simulation-intensive and the use of a hierarchical custom-design-package with an
interface to a powerful analogue simulator is required. (SL2000, HSPICE, STICKS). It is also
necessary to have access to test pattern generation and fault simulation software (HILO 3). For
practical ATE work it is also necessary to link an ATE system, such as a Tektronik DAS 9200, to
the CAD system in order that simulation test patterns can be automatically used to test the silicon
(HILO 3).

   The project plan is essentially the design and characterisation of cells which are then combined
to form complete ASIC backgrounds. Simulation time on the available packages is quite acceptable and enables multiple runs to be carried out. Following the design process it is intended that
suitable test structure will be fabricated using industrial prototyping and evaluated on the ATE
system.

2   *Optimisation of regular VLSI structures for silicon compilation (with GEC Hirst Research Ltd.,
Wembley)*
The main aim of the project is to produce new optimized algorithms and rules/strategies for
achieving a comprehensive functional-block-based silicon-compiler system. This should enable
system designers to have flexibility on important system parameters and is relatively technology-
independent.

   The project contains extensive circuit simulation work using HSPICE on regular structures such
as SRAM and the use of symbolic layout tool STICKS to achieve technology independence.

   The project plan initially concerns the use of circuit simulation to enable the dependency of
system parameters (speed, power, area etc) on circuit and technology considerations, to be
ascertained. Extraction techniques are then used to convert the algorithms into a suitable
mathematical form. These algorithms are then optimized using deterministic and statistical
techniques in order to achieve an optimal solution. It is expected that the system will be implemented in 'C'. Ultimately the work will be extended to cover other types of VLSI systems based on
regular structures such as PLA and systolic arrays.

3   *Design of high speed, high resolution data converters and associated voltage references (with
Plessey Semiconductors, Oldham)*
This project is aimed at extending the speed and resolution capabilities of analogue-digital

converters for use in digital signal processing applications such as video, high speed radar and thermal imaging. Inherent within this is the need for accurate, low temperature coefficient voltage-reference circuits.

In particular the project has concentrated on the design of 14 and 16 bit resolution DACs, a 200 MHz 8 bit ADC, and 2.5 and 5.0 V precision voltage reference devices.

The project required extensive circuit simulation work using HSPICE as well as the development of improved models for the BJT devices. Currently prototype versions of the devices are being fabricated and will be extensively characterised before the project is completed.

CONCLUSIONS

This note has attempted to show how the ECAD facilities are being used in collaborative post-graduate research at the Department of Electrical and Electronic Engineering at Huddersfield Polytechnic. In our view much of this work would not have been possible without the extensive range of design tools being made available under the ECAD Initiative.

EXPERIENCES OF SOLO 1000 IN RESEARCH AT THE UNIVERSITY OF EDINBURGH

M. S. McGREGOR, A. TOMLINSON and O. K. TAN
Department of Electrical Engineering, University of Edinburgh, Scotland

INTRODUCTION

The University of Edinburgh and European Silicon Structures (ES2) are jointly participating in ALVEY programme ARCH_020 'Advances in the Architecture and Compilation of Bit-serial VLSI Signal Processors'. One of the goals of this programme is to port the process-dependent cell library of the FIRST bit-serial silicon compiler to ES2's process independent silicon compiler SOLO 1000.

During the course of the programme Edinburgh has had significant experience of using SOLO 1000 on SUN workstations. These experiences are discussed via three of the major devices we have designed to date.

The data encryption chip

The encryption chip is a pin-compatible replacement for the industry-standard AMD DES processor Am9518 but which uses a different encryption algorithm. This algorithm was implemented using a 32 bit data path with three 32 bit buses and a PLA control structure. The data encryption device uses approximately 30K transistors most of which are used by 32 bit registers.

The radix 4 FFT butterfly chip

The butterfly chip is a bit-serial design using about 56K transistors. It multiplies four data words by four 16 bit coefficients and calculates a four point DFT of the product. The four serial/parallel complex multipliers use distributed arithmetic and offset binary coding to reduce the number of real multipliers needed from four to two. The four point DFT is then implemented using sixteen bit serial carry save adders to do two 2 point DFTs.

The speech synthesiser

The speech synthesiser is also a bit-serial design. It uses approximately 80K transistors. It is based on the acoustic theory of speech production. The device contains eight formant resonators (to simulate the vocal tract transfer function) arranged in parallel. Each formant channel has its own amplitude control and is excited by a combined voice and noise excitation. This allows the synthesis of both voiced and fricative speech. The outputs from each of these formant channels are then combined to obtain a single output for the speech waveform.

The above chips have been designed using SOLO 1000 over a 12 month period and the design systems advantages and disadvantages which emerged during that time are outlined below.

ADVANTAGES

(1)   The system seemed to be easy to learn and use. Designers with no experience of the toolset became familiar with its basic facilities within one week.

(2)   The silicon compilation produces fairly efficient layout in terms of area. Comparisons with other compilers (GENESIL, SECOND) available within the University have shown SOLO 1000 to often be as good as, or better than the alternatives.

(3)   The system is process-independent.

(4)   ES2 has a high success rate of designs submitted for fabrication.

(5)   Artwork can be generated very quickly (within one day for very large designs) once the chip is defined in the input language using either DRAFT (schematic capture) or MODEL (textual description).

DISADVANTAGES

(1)   Memory requirements: the system software occupies 11 mb of memory. A further 32 mb of swap space is needed if large designs are to be simulated.

(2)   For a subsystem of two 16-bit double-precision multipliers which has about 8,000 stages, run time for the simulation of 65 clock cycles is of the order of one hour real time on a SUN 3/110 with 4MB of internal ram and a floating point co-processor. As the design gets larger, run time increases drastically. Simulation of 200 clock cycles of one of the speech synthesiser devices (see point 7 below) takes five real days to complete when the SUN is used solely for that simulation. The simulator does not appear to handle its memory requirements very effectively. The above 200 clock cycle simulation required 20MB of virtual memory. Since the SUN uses about 2MB for overheads (e.g. operating system, networking etc) the user only gets about 2MB for his simulation. As a result, the user was only getting six percent of the cpu time for useful processing. Expansion of the SUN's memory by a further 8MB (effectively giving the user 10MB) reduced the simulation time by a factor of about nine and the device can now be simulated in just over 12 hours.

(3)   The simulator operates only at switch level, so for large designs the time taken to swap netlists in and out of the computer is far greater than the actual CPU time. Simulations can take a long time (> 24 hrs) to run. The system would benefit greatly from a functional simulator.

(4)   The schematic capture software (DRAFT) is not very robust, causing frequent crashes. Also it is a fairly inefficient means to enter designs into SOLO 1000. Direct entry via MODEL proves to be much quicker (but less well annotated).

The sheet size, once fixed, cannot be changed for plotting purposes. It would be useful to make the choice of sheet size available to the user during plotting. The choice of area to be plotted can also come in the form of a boundary definition to be specified during plotting.

There is no 'UNDO' facility to cancel previous commands. This can be particularly annoying as it is easy to fall foul of 'finger trouble'. E.g. we often end up deleting a whole wire net when we actually wanted to simply delete a line.

Moving instances will render the loss of wire connection to the instance moved. There are such facility in other schematic editors whereby connection of wires are intact even if the instance is moved.

Group of instances cannot be moved to an area which is overlapping the original area. That is, one cannot reposition the layout of the schematic diagram readily.

(5)   The lack of ROM/RAM/PLA cells within SOLO 1000 meant that on-chip memory had to be built from static latches and so took up a lot of silicon area. Control structures suffered from the same problem, being constructed from static gates.

(6)   The test vector input language is low level.

(7)   The maximum size imposed is about 32,000 stages (one stage is equivalent to one p-type and one n-type transistor) or 64,000 transistors. This limit proved to be too small for the speech synthesiser. It came ouy at about 40,000 stages. The device had to be split into two chips, each about of about 20,000 stages. This splitting up of the design proved to be very

problematical during simulation as only one device could be simulated at a time. We had to resort to other simulators (mostly FIRST) to verify the design.

(8)   The two-dimensional placement option can be useful for bit-parallel designs, but produces little improvement in area for bit-serial designs (which are naturally more wire conservative than bit-parallel designs). The speech synthesiser using one-dimensional array placement, occupies a total chip size of 42.89 mm^2, and has a run time of seven minutes. Using two dimensional placement the chip area was reduced to 42.07 mm^2, the run time however increased to over three hours of CPU time. Similar results were seen for the FFT butterfly, using 28,000 stages it consumed 69.6 mm^2 when constructed with one dimensional placement. In contrast, the data encryption device, using 15,000 stages, used 88 mm^2 when constructed with one dimensional placement. Two dimensional placement of this device reduced its area to 59.2 mm^2.

(9)   All of the input pads are non-inverting and all of the output pads are inverting. This requires all the output signals to be inverted first before feeding to the output pads. It would be useful if non-inverting output pads were also available.

CONCLUSION

SOLO 1000 is a reasonable design tool for systems up to 20K transistors which do not have a lot of on-chip memory or complex control structures, and do not require an excessive amount of test vectors. Larger ASICs, while do-able, become a problem as you quickly require much more processing power and memory to do short simulations. Also the maximum design size of about 64K transistors may be a problem for research work.

INDEX OF SOFTWARE PACKAGES USED

REDCAD (PCB design package)
Department of Electrical and Electronic Engineering, Huddersfield Polytechnic, 61

SILVAR LISCO — CASS (schematic capture package)
Department of Electrical and Electronic Engineering, The Queen's University of Belfast, 76
Department of Electrical and Electronic Engineering, University of Bristol, 92
Electronic Engineering Laboratories, University of Kent at Canterbury, 113

SILVAR LISCO — IHILO (HILO interface)
Department of Electrical and Electronic Engineering, The Queen's University of Belfast, 76
Department of Electronics and Computer Science, University of Southampton, 167

SILVAR LISCO — BIMOS (logic simulator)
Electronic Engineering Laboratories, University of Kent at Canterbury, 38

SILVAR LISCO — SWAP (switched capacitor simulator)
Department of Electronic Systems Engineering, University of Essex, 175

SILVAR LISCO — HELIX (mixed level simulator)
Department of Electronic Systems Engineering, University of Essex, 11
Department of Engineering Science, University of Exeter, 23
Electronic Engineering Laboratories, University of Kent at Canterbury, 38
Department of Physics, The University College of Wales, Aberystwyth, 52
Department of Electrical and Electronic Engineering, University of Bristol, 92
Department of Electrical and Electronic Engineering, The University of Newcastle upon Tyne, 146
Robert Gordon's Institute of Technology, Aberdeen, 156

SILVAR LISCO — GARDS (gate-array layout package)
Department of Electronic Systems Engineering, University of Essex, 11
Department of Physics, The University College of Wales, Aberystwyth, 52

SILVAR LISCO — CAL-MP (standard cell layout package)
Department of Electronics and Computer Science, University of Southampton, 167

SMS CELL LIBRARY (3 micron CMOS digital cell library)
Department of Electronics and Computer Science, University of Southampton, 167